Paleopathology
in Perspective

Paleopathology in Perspective

Bone Health and Disease through Time

Elizabeth Weiss

ROWMAN & LITTLEFIELD

Lanham • Boulder • New York • London

Published by Rowman & Littlefield
A wholly owned subsidiary of The Rowman & Littlefield Publishing Group, Inc.
4501 Forbes Boulevard, Suite 200, Lanham, Maryland 20706
www.rowman.com

Unit A, Whitacre Mews, 26-34 Stannary Street, London SE11 4AB, United Kingdom

British Library Cataloguing in Publication Information Available

Library of Congress Cataloging-in-Publication Data

Weiss, Elizabeth.
  Paleopathology in perspective : bone health and disease through time /
Elizabeth Weiss.
  pages cm
  Includes bibliographical references and index.
  ISBN 978-0-7591-2403-5 (cloth : alk. paper) — ISBN 978-0-7591-2442-4 (pbk. :
alk. paper) — ISBN 978-0-7591-2404-2 (electronic)
  1. Paleopathology. I. Title.
  R134.8.W46 2015
  616.07—dc23

                                                                    2014039779

∞™ The paper used in this publication meets the minimum requirements of
American National Standard for Information Sciences—Permanence of Paper
for Printed Library Materials, ANSI/NISO Z39.48-1992.

Printed in the United States of America

Dedicated to C.L.

Contents

Figures

Acknowledgments

WITHOUT THE HELP OF MANY INDIVIDUALS, THIS BOOK WOULD NEVER HAVE been written. I would initially like to thank Wendi Schnaufer who contacted me and met with me to discuss my book idea; her enthusiasm for the project enabled me to put together a book proposal that was approved by esteemed anthropologists Dr. Samantha Hens and Dr. Della Cook. Dr. Hens and Dr. Cook both provided me with excellent feedback that improved the book, including adding a chapter on dentition and providing a more thorough introduction to bone biology. Information on willed-body collections would have been far sparser without the generous assistance of Dr. Kate Spradely, Sophia Mavroudas, Dr. Allison Leitch, Dr. Robert Franciscus, and Shirley Schermer; they all provided me with quick responses regarding the most up-to-date information about the sex, age, and ethnicity of the collections and in some cases extensive historical information about the formation of the collections. I would also like to thank Leanne Silverman and Andrea Offdenkamp Kendrick at Rowman & Littlefield, both of whom helped in shaping the final version of the book.

I extend my gratitude to Dr. Charles Darrah of San José State University who has supported my academic endeavors since being on the hiring committee back in 2004. The entire anthropology faculty and staff have provided a productive and fun work environment which enabled my career to flourish from a traditional bioarchaeologist to an integrative anthropologist bringing together the past and present. Furthermore, I thank the San José State University Research, Scholarship, and Creative Activity Grant committee who were kind enough to grant me a course release to finish the book and financial support to hire students for the photos and illustrations. Students Daniel Salcedo and Vanessa Corrales deserve special mention for their dedication and the beautiful jobs they did on the book's images, which enhance the end product greatly.

On a personal note, I would like to acknowledge my family and husband. I thank my siblings Katherine, Alex, and Chris, who all have been continuously supportive of my academic career. I thank Nick Pope, my husband, for doing the shopping, cooking, and cleaning while I was on campus and hard at work. I also need to thank Nick for proofreading this book; without his help, children would have more caries as a result of snaking rather than snacking. I also thank my relatives Dr. Jutta Brederhoff and Joachim Leisegang for their continuous interest in my success. Finally, very special thanks go to my parents Gisela and David; they raised four successful offspring and made it seem effortless. They have always been there for me and yet raised me to be remarkably independent; I can never thank them enough.

CHAPTER 1

Introduction to Bone Research

OSTEOLOGY IS THE STUDY OF BONES. ANTHROPOLOGISTS WHO UTILIZE knowledge about bones include bioarchaeologists who study human remains in the archaeological record, forensic anthropologists who study skeletal remains in a legal setting (such as when a crime has been committed, to aid in victim identification), paleoanthropologists who examine the fossil record to reconstruct the lives of early human ancestors, biological anthropologists who may study the effects of pharmaceutical drugs on bone biology, and **anthropometric** anthropologists who may aid in determining ergonomic workplaces. A firm grasp of understanding osteology is essential for all of these anthropologists, but individuals involved in the health sciences, such as medical doctors specializing in orthopedics and physical therapists engaged in rehabilitation, also require extensive knowledge about bone biology.

BONE BIOLOGY

To understand how skeletal remains can be indicative of health, one needs to understand bone biology. Bones have multiple functions. Bones protect organs, such as the lungs residing in the torso that are protected by the ribs. Bones also support soft tissue, including skin. Bones furnish surfaces for muscles, **tendons**, **ligaments**, and cartilage to attach onto. With these connections, the skeleton acts as a lever system that enables us to move (White and Folkens 1991). Bones provide fat storage in the **medullary cavity** of long bone shafts and between bone cells in **trabecular bone**. Bones are also the site for production of the blood cells that fight off infections. Maintaining **calcium** and **phosphorus homeostasis** is another function of bone. Without this maintenance, hypocalcemia (i.e., levels of calcium that are too low) can occur and cause health problems, such as muscle spasms, heart dysfunctions, and seizures; or hypercalcemia (i.e., levels of calcium that are too high) can arise and cause widespread organ dysfunction and damage (Confavreux 2011).

Bone tissue is dynamic. It changes with the growth of an individual, and it interacts with **stresses** placed on it by mechanical loading, such as body weight and muscle use (White and Folkens 1991). Bone is made up of organic components, such as **collagen**, and inorganic components, such as calcium. Collagen provides bone with ductile properties that allow bone to warp rather than break when stressed; the **mineral** constituents of bone account for brittleness and provide strength to the skeletal system.

At the microscopic level, osteologists refer to bone as being either immature or mature. Immature bone is found in infants, in children, and after trauma has occurred in individuals of all ages. Immature bone is disorganized and not sequenced. It also has a greater amount of bone cells than mature bone. Mature bone replaces immature bone once the growth or healing has ceased. Mature bone is arranged in the Haversian system. Due to the Haversian system's **anisotropic** (i.e., unequal physical properties along different axes) quality, bone responds well to stresses and strains without remodeling in the long axis, but bone requires remodeling to avoid breakage in the horizontal axis (Hamill and Knutzen 1995). In other words, bone is ideally organized to bear weight. The Haversian system consists of a central canal that is surrounded by hard calcified matrices called **lamellae**, small spaces with mature bone cells called **lacunae**, and small canals called **canaliculi** (Hamill and Knutzen 1995). As illustrated in figure 1.1, the central canals run through bone longitudinally (along its longest axis) and are surrounded by lamellae; between the lamellae

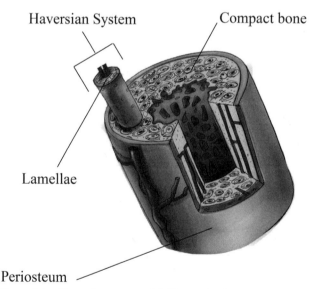

Haversian System   Compact bone

Lamellae

Periosteum

Figure 1.1.   Haversian system. This illustration highlights the organized nature of mature bone. Illustration by Vanessa Corrales.

are lacunae, and radiating from these lacunae are canaliculi that are filled with phosphate ions and calcium (White and Folkens 1991). Canaliculi connect lacunae with each other and with the central canals. The Haversian system provides routes for nutrients, bone and blood cells, and oxygen to travel, which enables bone to stay alive.

At the macroscopic level, adult skeletons consist of cortical or compact bone and trabecular or spongy bone. **Cortical bone** is solid and dense. Eighty percent of the adult human skeleton is made of cortical bone, and 20% consists of trabecular bone (Liebschner 2004). Long bone shafts, which are called **diaphyses**, consist almost entirely of cortical bone. All other bones, such as cranial bones, ribs, and vertebrae, contain a thin layer of cortical bone. Cortical bone is the most abundant type of bone in bioarchaeological sites since its denseness aids in preservation. Trabecular bone is porous, which reduces the chance of its preservation. Although trabecular bone is less likely to preserve due to its architecture, trabecular bone has essentially the same components as cortical bone (Keaveny et al. 2001; Liebschner 2004). Trabecular bone is arranged in packets of lamellae rather than in the Haversian canal system; it is more heterogeneous in its three-dimensional arrangement and tissue properties. Essentially, within an area of trabecular bone the direction of the lamellar packets and the density of these packets vary depending on the direction of the stresses (Keaveny et al. 2001). Trabecular bone is found at the end of long bones, which are called the **epiphyses**. During growth, epiphyses are unfused and apart from the shaft. Between the shaft and the epiphyses are cartilaginous growth plates called the **epiphyseal plates**; they enable growth in length (rather than in diameter), and once they fuse, growth in length ceases (White and Folkens 1991).

During life all bone is covered with two tissues: the **periosteum** on the outside and the **endosteum** on the inside. These tissues are bone forming or osteogenic. Bone-forming cells, which reside in the endosteum and the periosteum, are especially abundant in growing individuals, but bone-forming cells are active and present throughout life to replace, repair, and remodel bone (White and Folkens 1991). For adults, bone-forming tissues are most active during trauma, such as when a bone is broken. Cells at the periosteum are more active during adulthood; whereas the cells at the endosteum are more active in younger individuals. Thus, old individuals' bones are rough compared to young individuals' bones (figure 1.2).

Osteogenesis, the creation of bone, occurs during repair and remodeling; the basic components of these biological functions are the same. Bone remodeling, like bone repair, occurs at particular locations due to muscle use and other stresses. Determining the difference between repairing and remodeling is difficult due to methodological shortcomings in various studies. For example, some studies use vises inserted into bone tissue to measure **strain**

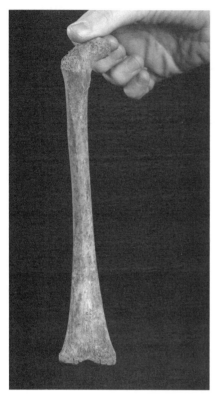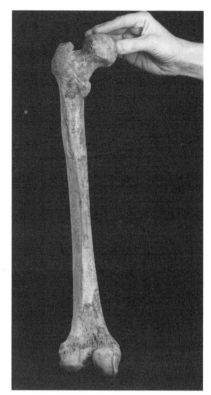

Figure 1.2. Young and old bones. On the left side is a smooth femur from a five- to seven-year-old, whereas the right side displays a rough femur from a thirty-five- to forty-five-year-old. Photographs by Daniel Salcedo.

and remodeling. Even studies that do not use vises still use X-rays to calculate remodeling, and the radiation in X-rays may alter the behavior of bone (Currey 2012). The first person to provide evidence for bone remodeling was German anatomist Julius Wolff in 1892, whose statement on it became known as **Wolff's law**. His law can be summarized as follows: the internal architecture and external form of bone alters following mathematical rules when subjected to strains, which are external forces, and stresses, which are internal forces (Wolff 1986). Wolff's law indicates two different types of reactions to forces on bone: bone remodeling that increases bone girth at cortical bone locations and reorganization of trabecular bone (Keaveny et al. 2001). Dynamic strain is necessary to initiate bone remodeling (Mosley 2000). Greater levels of strain increase bone remodeling, but excessive strain can cause deformation of bones and may cause fractures; determining how much strain is too much depends on an individual's age, the location of the strain, overall health, and the type of force that causes the strain (Keaveny et al. 2001).

Osteocytes are bone cells. Osteocytes are the most abundant cells in bone; they are long living and were once thought to be static, but now osteologists understand that they respond to stimuli and initiate bone remodeling (Heino et al. 2009; Neve et al. 2012). Bone remodeling requires a lot of energy and is controlled by **hormones**, such as **leptin**, serotonin, and osteocalcin (Mosley 2000). The parathyroid hormones are very important in the regulation of calcium homeostasis (Hadjidakis and Androulakis 2006). Many of these hormones also regulate **metabolism** and appetite. Due to the requirement of strain and the importance of hormones, such as leptins that are found in fat cells, obesity has been suggested to protect against osteoporosis (Mosley 2000).

There are two types of osteocytes: **osteoclasts** and **osteoblasts**. These cells are responsible for remodeling and repairing bone (White and Folkens 1991). Osteoclasts develop from white blood cells and resorb (or take away) bone; osteoclasts' action of resorption leaves a lacuna where osteoblasts deposit precalcified bone matrix (Confavreux 2011). **Bone resorption** occurs continuously throughout life and outpaces bone depositing later in life. Osteoblasts are responsible for laying down new bone material. Osteoblasts make uncalcified pre-bone tissue known as **osteoids**. The last part in both bone repair and remodeling is **calcification**. Calcification enables bone to become hard. Without calcification, bones cannot bear weight. About 90% of calcium is stored in bone where it can be reabsorbed by blood and tissue. Still, calcium consumption is needed for bone development and maintenance. **Fluoride** is another mineral important for bones, especially for growth and protection against **demineralization** of bone. Fluoride is usually added to toothpaste and water to prevent demineralization of enamel. Vitamin D regulates calcium absorption; without vitamin D, the body cannot absorb calcium. Vitamin D is a hormone that resides in the skin that is activated through exposure to ultraviolet radiation. There are dietary sources of vitamin D, such as fish and eggs, but the amount of vitamin D in these foods varies depending on the amount of sun exposure these foods experienced. In other words, fish that swam in waters where little sun exposure occurred have low levels of vitamin D.

TEMPORAL CHANGES IN HUMAN LIFESTYLE

Whether looking at past or present populations, skeletal remains can reveal information about dietary deficiencies, activity levels, infections, and traumas. Human lives have changed drastically over the last two hundred thousand years when *Homo sapiens* first evolved from *Homo heidelbergensis*. Some of these changes will be detectable in the skeletal records since they are accompanied by changes in health, activity, and trauma. Changes that occurred over long periods of time as a result of evolution by natural selection may also be observable in skeletal samples; for example, changes in limb proportions as a result of climate adaptation can be measured through skeletal remains. **Allen's rule**, which contends that shorter distal elements will be found in cold-adapted species, is a good example of climate adaptation. For instance, Asian populations who are cold adapted have short shinbones in relation to their thighbones, but warm-adapted African populations have long shinbones in relation to their thighbones. The invention of tools, such as bows and arrows, that allowed humans to hunt more effectively also reduced trauma. It appears that Neanderthals with their lack of throwing weapons were more prone to trauma when hunting large animals than were *Homo sapiens* who were able to distance themselves from their prey (Trinkaus and Zimmerman 1982). The building of semi-permanent or permanent housing encouraged settling down rather than continuing a nomadic lifestyle; these changes occurred during the **Neolithic Revolution** around ten thousand years ago in the **Levant**. As found in the bioarchaeological record, a reduction in roaming has led to a decrease in lower limb strength in many populations (e.g., Bridges 1989; Kimura 1984; Lieverse et al. 2007).

In the last ten thousand years, significant changes have occurred, such as the adoption of agriculture, the domestication of animals, and immense population growth that led to over-crowding, decreases in sanitation, and an increase in infections. Evidence of the increase in infections related to population growth is abundant in bioarchaeological skeletal samples (e.g., Hutchinson and Richman 2006; Oxenham et al. 2005; Suzuki and Inoue 2007). Nonetheless, agriculture was not universally adopted. Places such as California, where the natural abundance of resources year-round enabled population growth without agriculture, did not change their subsistence patterns until very recently.

After agriculture was widely adopted and modified into a variety of forms, including pastoralism and horticulturalism, it took nearly eight thousand years for another major transition to occur; the **Industrial Age** arose in the last couple of hundred years. The changes that occurred in the Industrial Age included a vast movement into urban areas, especially in Europe. There was a continuation of over-crowding and poor sanitary conditions. Infectious diseases, such as tuberculosis and the bubonic plague, flourished during this period and can be seen in examinations of historical skeletal samples from Europe (e.g., Djurić-Srejić and Roberts 2001). The lethality of the diseases may have been exacerbated by the poor conditions of those living in urban areas; children, who were especially vulnerable, worked long hours and faced severe malnutrition. Furthermore, the smog in cities such as London coupled with naturally cloudy environments and long hours spent working in factories led to **epidemics** of vitamin D deficiency and the resultant bone disease **rick-**

ets. Although evidence of vitamin D deficiency is rare in prehistoric skeletal samples, its manifestations are abundant in skeletal remains from the 1800s in Europe (Brickley et al. 2005; Mays et al. 2009). Although the Industrial Revolution was not prevalent in Africa, Central and South America, and a variety of other locations, recent **urbanization** was born out of the Industrial Revolution. And although there are still populations that practice hunting and gathering or a mix of foraging, hunting, and horticulture, over half of the world's population now lives in cities. Even the most iconic of the hunter-gatherer populations, such as the San of South Africa, have replaced their subsistence pattern in the twentieth century, with fewer than 5% of their population existing on foraged and hunted meals.

Recently, changes in the environment and culture have led to further changes in bone health. Even after child labor laws enabled children to spend their hours outside of factories and after technical advances enabled cleaner environments, changes continue that affect human health. In the last half century, changes that affect bone health have not slowed down. Computers have increased the number of **sedentary** jobs, increases in high-calorie food and drink have fueled the obesity epidemic, and yet longevity has increased throughout the Western world as a result of medical care. Changes that have occurred in the last fifty or sixty years that may affect health as observed in bones include an older population, increased weight, increased diversity, and decreased inbreeding. Children's sports have become more intense lately, while other children's lives are marked by an increase in sedentary lifestyle united with high-caloric diets. Also, hormonal exposure in the young and the old has increased through the last forty or fifty years, which could affect development, growth, and bone loss. Most of these changes' effects on bone have not been as well studied using a comparative method as have the changes in the earlier part of prehistory and history, but these changes may have led to increases in back ailments, osteoarthritis, and **osteoporosis**, which are all health issues that can be assessed through the study of skeletal remains. Although these changes are perhaps most pronounced in Europe and the United States, similar trends are found in Africa, Asia, Australia, and South America. Yet there are always exceptions, such as the Shuar of Ecuador. This book addresses changes that leave visible marks on bones that have occurred in the last six decades, with a focus on global trends. It will describe how these changes, which are observable in skeletons, have affected human health.

EVIDENCE OF CHANGE: SKELETAL SAMPLES AND CLINICAL DATABASES

Bones, as will be demonstrated throughout this book, are excellent forms of data because they can reveal changes in health without written records. Osteologists can determine demographic changes in degenerative diseases, such as osteoarthritis and osteoporosis, with a good deal of certainty. Osteologists can also draw conclusions about inbreeding, **congenital** diseases, rheumatoid **arthritis**, malnutrition, infectious diseases, and many other health factors.

Although the osteological record reveals much about health, it is not a perfect record. For one, many diseases leave similar marks on bones; two good examples of this dilemma include similarities between tuberculosis compared to bronchitis and similarities between venereal syphilis compared to non-venereal syphilis. Additionally, many health indicators, with the exception of degenerative diseases, are indicators of childhood health rather than adult health. Poor-health indicators on bone can also suggest that individuals were actually fairly healthy since most diseases, such as tuberculosis, affect bones last, and thus those individuals with the weakest immune systems died before the bones were affected. This conundrum is an essential part of the osteological paradox, which was first raised to explore whether the transition to agriculture from hunting and gathering improved or deteriorated health (Wood et al. 1992). A child's skeleton that looks completely healthy and has no injuries is still evidence of a child who died before reaching adulthood, and therefore the child was likely ill. An additional problem with the bioarchaeological record is that many diseases do not affect bone. Anthropologists do not know how past populations were affected by arteriosclerosis, **cardiovascular disease**, strokes, and cancers because adequate data are not available. Soft tissues, like arteries and muscles, are unlikely to preserve. Finally, bone remodels, and this can erase the evidence of health stresses if the individual recovered from his or her illness.

The types of data reviewed here to understand bone biology and health include animal experimental samples, bioarchaeological skeletal samples, historic skeletal samples, autopsy skeletal samples, and clinical data sets. For the most part, animal experimental models have been used to understand **etiology** of various diseases, such as osteoarthritis, and the biology involved in bone remodeling and repair. Animal models have been used to understand the effects of forces on bone, such as Woo et al.'s 1981 work on swine that demonstrated bones increased in strength after exercise or Zumwalt's 2006 work on sheep where she found that exercise did not alter muscle attachment sites located on various long bones. Benefits of animal models are that you can control the animal's environment and then sacrifice the animal to see the effect the environment had on the bones. Conversely, non-human animal bone also differs slightly from human bone; for example, rodents tend to have fine fibrous bone tissue and no intracortical remodeling as humans do. Dogs lack lamellar structures, and their bone consists of woven fibers (Liebschner 2004). Whether these differences affect the understanding of human disease remains unknown.

Bioarchaeological skeletal samples are good for comparative studies of changes that have occurred over thousands of years. There are thousands of bioarchaeological skeletal samples. The samples range from large samples, such as CA-Ala-329, a hunter-gatherer collection of nearly 300 individuals, to single individuals, such as Kennewick Man. The remains come from both the **New World** (North America and South America) and the **Old World** (Europe and Asia mainly, with few remains from Africa). Hunter-gatherer-fishers and agriculturalists are represented. Preservation quality varies greatly.

Bioarchaeological skeletal collections, unfortunately, lack medical records, and as a consequence diagnosing diseases or determining quality of life may be difficult and in some cases impossible. Research on taphonomy, which has enabled anthropologists to understand what happens to remains from the time they are buried until their discovery, has demonstrated that burial samples are also not demographically similar to living populations. For instance, young individuals do not preserve as well as older individuals. Individuals who have not finished growing have a greater number of bones prior to the fusion of the epiphyseal plates, and many of these unfused bones are small and thus are likely to be destroyed or missed during excavation. Infant remains are particularly rare since their bones have yet to completely calcify in locations, such as the skull. Plus, infant bones are small. Even within an individual, certain parts of the skeleton do not preserve as well as other parts. Ends of long bones that are made up of trabecular bone do not preserve as well as the shafts of long bones that consist of cortical bone. And small bones, such as hand bones, are often lost during the excavation. Old individuals are absent in many bioarchaeological samples due to the shorter lifespan of past populations; in CA-Ala-329, for instance, the average age of death was thirty-five years of age. A skewed age demographic may make drawing conclusions about degenerative diseases difficult. Another dilemma is that some of the samples are created from remains that spanned hundreds and even thousands of years, which is the case with CA-Ala-329. Anthropologists have questioned whether it is valid to consider a sample that spanned several thousand years as one population, whereas in autopsy collections the span tends to be less than one hundred years.

Historical samples tend to come from European cemeteries that were excavated, but a few are from the Americas, such as the Quebec prisoners of war housed at the Canadian Museum of Civilization in Hull, Quebec. Historical samples are better than bioarchaeological samples in some respects. Historical samples often have greater preservation due to burial practices compared to bioarchaeological samples. Gravestones often indicate the sex and age of the individual. There also may be written records or accounts of the area accompanying the graves that may help determine causes of death. The Quebec prisoners of war had food diaries, activity logs, and pre-military enlistment occupation records. Nevertheless, overall historic samples tend not to represent the whole population of the time period; they may over-represent specific peoples, such as clergymen, the wealthy who could afford a funeral, military personnel, slaves, or hospital patients.

Autopsy or documented skeletal collections coupled with medical databases, such as those from the National Institutes of Health (NIH), are best for comparisons of changes in the last century to half century. There are nine documented skeletal collections in the United States. Five of the nine collections have ceased growing, which means the institutions that curate these collections are no longer accepting bodies. The samples that have ceased growing are representative of people who died before the 1970s. For the autopsy collections, information regarding sex, age, and occupation is often available. Sometimes health records and cause of death have been obtained, but information on hobbies, diet, and other details is often unknown, especially in the oldest collections. Autopsy collections consist mostly of indigent individuals who did not leave money for a funeral or who were not claimed by relatives. Many of the remains come from individuals who were homeless; these individuals during their lives would have been at greater risk of infections, such as tuberculosis, and other communicable diseases. Their alcohol consumption was likely high, whereas their diets were likely poorer than the population they are representing. Plus, the samples have more males than females, contain few children, and consist mainly of whites and blacks.

The Smithsonian's National Museum of Natural History houses two autopsy collections: the Robert J. Terry Collection (1871–1966) and the Huntington Collection (1892–1920). Unless otherwise noted, information on the Smithsonian's National Museum of Natural History collections comes from their museum's website (SINMNH 2013). Dr. Robert J. Terry was an anatomy professor at Washington University in St. Louis, Missouri, from 1899 to 1941. Terry had an intense interest in human skeletal anatomy and **pathological** variations compared to normal variations found in bone. His mentor Dr. George S. Huntington (1861–1927), who was a professor of anatomy at Columbia University's College of Physicians and Surgeons in New York for thirty-five years, impressed upon him the importance of preserving cadavers and skeletons for research (Hrdlička 1937). Huntington developed his own collection of human remains gathered from European immigrants and New York City residents for his research (Hrdlička 1937). The Huntington Collection contains 3,600 individuals of known sex, age, nationality, and cause of death collected between 1892 and 1920, but this collection consists only of crania. Nevertheless, there are examples of trauma, infections, tumors, and congenital defects in the Huntington Collection that have been of great use to osteologists.

Terry obtained skeletal remains from anatomy classes and institutional morgues in Missouri. In the 1920s, Terry started his work on the collection and perfected a method of preserving the skeletons that would not remove all the fat in the bones. Having a bit of the fat left in the bone, Terry predicted correctly, would prevent them from becoming brittle and fragile. For the Terry Collection, human remains were collected from 1927 to 1967. In the early part of the collection, most remains were from low-income or institutionalized males whose bodies had not been claimed, but in 1955–1956 a state law requiring body donation to be willed ended this practice. On the retirement of Terry in 1941, Dr. Mildred Trotter (1899–1991) took over the curation of the collection. Under her supervision, the focus was on adding females and younger individuals to create a sample more representative of living populations. Most of the bodies during Trotter's time curating the collection were willed, and some had written informal medical histories. These medical histories have since been lost. In the late 1960s, Trotter was contemplating retirement, and the collection was handed over to Trotter's longtime friend and colleague Dr. Thomas Dale Stewart (1901–1997), who at the time worked at the Smithsonian (Pace 1997). The Terry Collection seemed to be in better hands under the management of Stewart since Washington University had shifted its focus to brain morphology and function (Pace 1997). In 1967, the collection was moved to the Smithsonian's National Museum of Natural History. Although about 2,000 individuals had been processed, the collection currently has a total of 1,728 individuals. Some individuals were sent to other collections, such as those sent to Dr. Raymond Dart at the University of the Witwatersrand in South Africa. There are 461 white males, 546 black males, 323 white females, 392 black females, and 5 Asian males. The birth dates of the individuals ranged from 1822 to 1943. There are only fifty-three individuals who were under twenty-one years of age at death and only fourteen individuals over the age of ninety. For about 60% of the collection, anthropometric data and cadaver photographs are available. For some individuals, dental charts, death masks, and cause of death are available.

Another impressive collection is the Hamann-Todd Collection that was started at Western Reserve University (which is now known as Case Western Reserve University) in Cleveland, Ohio, after a 1911 state law allowed for the preservation and curation of human remains. Information on the Hamann-Todd Collection comes from the Cleveland Museum of Natural History's website (CMNH 2013). Dr. Thomas Wingate Todd (1885–1938) replaced Dr. Carl Hamann (1868–1930) as professor of anatomy at the Western Reserve University medical school. Todd started the Hamann-Todd Collection, so named due to the support Hamann, who had become dean of Western Reserve University, provided Todd in building his collection. By the time of Todd's death in 1938, 3,600 individuals were collected, which made the Hamann-Todd Collection the largest autopsy collection in the world. Sex, age, ethnicity, weight, and height were collected for each individual. During the 1950s to 1960s, the collection was temporarily transferred to the Cleveland Natural History Museum. The usefulness of the collection, especially in terms of comparative anatomy with early hominids, led the museum's curators to seek permanent curation rights for the collection. By 1970, the collection had been permanently transferred to the museum, where it remains to this day. The Hamann-Todd Collection contains 3,715 individuals, 6 of whom are Asian. There are 376 black females, 1,038 black males, 320 white females, and 1,932 white males. The mean age for the males is fifty-two years old, and for females it is only forty-four. Mean body mass index (BMI), which is a weight-to-height ratio, for both males and females was between eighteen and nineteen. To put these numbers in perspective, consider that according to the World Health Organization (WHO) in 2011, the mean age for US males at death was seventy-six years old while for females it was eighty-one. Moreover, according the WHO, the average BMI for males and females in 2008 was around twenty-eight.

The W. Montague Cobb Collection housed at Howard University in Washington, D.C., has been reported to be of similar quality to the Hamann-Todd Collection. Unless otherwise noted, information on the W. Montague Cobb Collection comes from the Howard University College of Sociology and Anthropology website on program resources (HUCSA 2013). The collection contains 732 individuals who occupied the lowest **socioeconomic statuses** in Washington, D.C., from the 1850s through the 1960s. The collection is named after Dr. William Montague Cobb (1904–1990), who obtained a medical degree from Howard University in 1929; he went on to obtain a PhD in physical anthropology from Western Reserve University. Todd was Cobb's mentor and taught him the importance of skeletal collections (Rankin-Hill and Blakey 1994). After completing his PhD in 1932, Cobb returned to Howard University to teach. He started a skeletal collection that he utilized in research that included debunking "race" differences in skeletal anatomy that provide blacks with an innate edge in running (Rankin-Hill and Blakey 1994). Cobb also studied the effects of racism on the health of African-Americans. More recently, Dr. Michael Blakely has worked on revitalizing the curation of the W. Montague Cobb Collection. Between 1992 and 1994, the curational facilities were updated and the information on individuals' sex, age, and ethnicity was digitized (Rankin-Hill and Blakey 1994).

One of the most interesting autopsy collections is from Stanford University, but it is now housed at the Office of the State Archaeologist in Iowa. In the early 1900s to the late 1950s, the Department of Anatomy at Stanford University Medical School collected skeletal remains. Dr. Arthur W. Meyer (1873–1966) was pivotal in creating the Stanford Collection; Meyer taught anatomy and emphasized the importance of having a skeletal collection for teaching and research (Clifton et al., n.d.). Meyer's focus was on pathologies, but non-pathological skeletal remains were collected for comparative purposes. In 1938, Meyer retired from teaching after twenty-nine years at Stanford (Clifton et al., n.d.). In the early 1980s, Stanford was short of space and looking to remove the pathological part of the collection to another location; they initially loaned the pathological individuals of the collection to San Diego's Museum of Man for an exhibit. The loan became permanent, and the Museum of Man still curates this portion of the collection. Stanford's active management of the collection ceased in 1959; the collection was largely inaccessible, except to professors who taught there such as Dr. Robert Franciscus. When Franciscus taught osteology and evolutionary anatomy at Stanford's Anthropology Department in 1995, the collection had been rummaged through over the years, yet the majority of the sample was still intact. Dr. Franciscus applied for a grant to improve the collection's storage and cataloging after being encouraged by the anatomy department to take control of the curation of the collection. In 1998, Franciscus joined the University of Iowa, and the collection was moved to Iowa. There are 1,100 individuals; information on age, sex, race, cause of death, marital status, occupation, place of birth, and medical history is available for some of the individuals (Schermer 2013). Birth dates range from 1845 to 1915. The individuals were residents of San Francisco, California, and thus there is greater ethnic diversity in this collection than in the Hamann-Todd Collection. The sex ratio is four males per one female, and the age at death ranges from twenty-seven to ninety-six years old (Schermer 2013).

The collections mentioned above are excellent for forensic research, and they are essential to understanding the changes in human health over the last fifty years, but they fail to accurately represent modern populations. Beyond the high rate of indigent individuals and the unrepresentative sex and ethnic distribution, there are also issues because of changes that have occurred over the last fifty years. For instance, life expectancy has increased from sixty-five years in 1950 to seventy-seven years in 2010. According to the US Census between 2000 and 2010, the population sixty-five years and over increased at a faster rate than the total US population. Additionally, the obesity rate has nearly tripled from 13% in 1960 to 34% in 2007. In females, hormonal birth control use during reproductive years has been linked to higher bone mass density in post-reproductive years in some studies and the opposite in other studies. **Oral contraceptives** were not even available in 1959, and now 30% of women of childbearing age in the United States use the pill. These changes alone may affect osteoarthritis and osteoporosis rates. Changes in lifestyle, such as rates of prescription medicine use, activity levels, dietary shifts, and smoking rates, and the consequences of these changes on bone health, make it evident that the populations in these autopsy samples are not comparable to present-day populations. Plus, the United States is growing more diverse; for example, the 2000 to 2010 census data reveals that the Asian and Hispanic populations both increased by over 40%. Different populations have different medical risks. For example, Caucasian-Americans are more likely to use only alternative medicine, such as **acupuncture**, than are African-Americans (Yang et al. 2012a). How many of the differences in bone health relate to lifestyle and culture and how many relate to biology can be addressed in diverse skeletal samples with good medical histories. To remedy some of the discrepancies between autopsy collections and living populations, some institutions have created new donated-body collections.

There are four active documented skeletal collections in the United States. The institutions involved in creating these collections accept bodies that anatomy courses may not, such as obese individuals or individuals with missing limbs. In all cases, individuals with HIV/AIDS, hepatitis, tuberculosis, and antibiotic resistant staph infections (MRSA) are not accepted; some of the collections mention other exclusionary diseases, such as Ebola. The four active documented skeletal collections are donated- or willed-body collections; that is, prior to their deaths, individuals decide to donate their bodies to the collections and fill out forms to that effect, or next of kin may decide to donate a body to one of these collections. The minority of individuals come from medical examiner offices, anatomy classes, and institutions (unlike in earlier autopsy collections). The benefit of the willed-body donation is that much about the individuals can be learned from the individuals themselves prior to their death. Forms for donation include questions regarding occupation, activities, health, weight, birth information, habits, socioeconomic status, sex, and birth date. After death, curators of these collections take a variety of anthropometric measurements and enter the data into FORDISC, which is a computer database used mainly for forensic anthropologists to help conduct research and apply the data to identifying victims of crime.

The smallest of these new collections, with seventy-five individuals (fifty-five whites, sixteen blacks, three Hispanics, and one of mixed heritage) as of 2014, is the Hamilton County Forensic Donated Collection located in Chattanooga, Tennessee. Due to the emphasis on forensic research and application, the information gathered for each individual has been limited to age, sex, race, and cause and manner of death. To avoid an over-representation of elderly white individuals, Allison Leitch (2012), the forensic technical specialist curating the collection, has stated that currently only non-white and young individuals are being accepted.

The largest of the continuously growing collections is the W. M. Bass Collection in Knoxville, Tennessee, which was started in 1981 and is named after longtime University of Tennessee forensic anthropology professor Dr. William M. Bass. Information from the W. M. Bass Collection was obtained from the University of Tennessee at Knoxville Forensic Anthropology Center website (UTK 2013). This collection has over one thousand individuals. As of 2014, the birth range of the sample is currently from 1892 to 2011; most individuals were born after 1940. There are forty-two infant and fetal remains. The skeletons come from thirty-six different states, but the majority are from Tennessee and the southeastern part of the United States. Nearly 90% of the individuals are white, and there are over twice as many males as females in the sample. This skewed demographic may change over time, and it is encouraging that about one hundred individuals are added to the collection each year.

In the Southwest, there are two willed-body collections: New Mexico's Maxwell Collection and the Forensic Anthropology Center Texas State (FACTS) Collection. Information regarding the Maxwell Collection was collected from the University of New Mexico Laboratory of Human Osteology website (UNM 2013). The Maxwell Collection is housed at the University of New Mexico in Albuquerque; currently it has 258 individuals, 75% of which are white. Both sexes are represented, as are all ages and many ethnicities. The Maxwell Collection was started in 1984, but it was not until 1995 that donors were asked to fill out information regarding occupation, health, weight, and activities.

The FACTS Collection was started as a willed-body donation collection in 2008; it is managed through Texas State University in San Marcos (FACTS 2013). So far, there are ninety-nine individuals in the FACTS Collection. The individuals have all been obtained from a two-hundred-mile radius, with San Marcos, Texas, being the center (FACTS 2013). There are thirty-seven females and sixty-two males (Mavroudas 2013). The majority of the individuals (eighty-six out of ninety-nine) are white. FACTS receives up to four donations a month; most of the individuals are low income, and they have opted to donate their body to avoid funeral costs (Martinez and Brunson 2012).

Clinical databases are usually large comparative samples created to address specific concerns. The databases are dependent on people who either volunteer due to concerns about their health or who volunteer for altruistic purposes. A good example of a clinical sample is the NIH Osteoarthritis Initiative database. The NIH Osteoarthritis Initiative enrolled 4,769 males and females between forty-five and seventy-nine years old at the onset of the four-year study period that began in 2004. Data were collected at four clinical sites (Baltimore, Maryland; Columbus, Ohio; Pittsburgh, Pennsylvania; and Pawtucket, Rhode Island) where it was possible to get an ethnically diverse sample. Another such database includes the Framingham Heart Study, which was started in 1948 in

Massachusetts under the direction of the NIH. The Framingham Heart Study initially recruited over five thousand men and women between the ages of thirty and sixty-two, but second and third generations of the initial cohort have been added to the database. This familial focus has allowed for genetic research on disease. Other databases may use data on averages provided by doctors or other health care workers. The WHO database on BMI, for example, has a set of criteria for accepting data that includes use of the international BMI cutoffs, means and standard deviations for the data, sex-segregated information, a minimum sample size of one hundred individuals, and standard measuring techniques described in their technical reports.

Clinical databases are flawed in that for many of the specific topic databases, sick individuals tend to be over-represented. Plus, the mean age of clinical databases tends to be greater than the general population, especially when the focus is on degenerative diseases. Nevertheless, the databases provide in-depth information on bone health, quality of life, family history, and lifestyle that may affect bone health.

Databases that have emerged from census collection are useful since they give researchers information about demographics (i.e., ethnicity, sex, and age) that can help anthropologists draw conclusions about bone health issues that may have increased lately. Using databases along with the skeletal samples mentioned above allows for comparisons to be drawn determining changes in the last half century.

Any one type of data is imperfect; to understand the changes of the last fifty to sixty years, data from bioarchaeological collections, historical skeletal remains, autopsy collections, and clinical databases are all needed. In this book, each topic will be examined from multiple data perspectives. However, one must consider that different diagnostic methods are used in clinical studies and osteological studies. This complicates comparisons but does not invalidate them. Throughout this book, different methodologies will be reviewed. The chapters are organized by bone trait categories. For each chapter, you will find an explanation of the traits, a review of the data from past and present populations, a description of how trends in traits have changed in the last fifty to sixty years, and the likely causes of these changes. By addressing the topics in this manner, I hope to show how human biology and behavior are intrinsically intertwined and that lifestyle impacts health.

CHAPTER 2

Growth Patterns

GROWTH IS A PERIOD WHERE THERE IS AN INCREASE IN THE NUMBER OF CELLS or in size. Studies on human growth have shown that skeletal growth is complicated. Skeletal growth is **polygenic** (i.e., controlled by multiple genes), a result of the effects of various hormones, and sensitive to the environment. Between 70% and 90% of variation in skeletal growth can be attributed to genetic differences (Silventoinen et al. 2011). Nevertheless, **environmental** stresses may halt growth or increase the growth rate. The earlier in life, the more flexibility there is in growth; for example, birth length is more greatly affected by prenatal environment rather than genes, and adult **stature** is more determined by genes than by environmental factors (Silventoinen et al. 2011). Part of the reason for this variable effect of the environment is that different genes will be active at different times in one's life; this is especially true of changes that occur as a result of **puberty** (Silventoinen et al. 2011). Moreover, twin studies have revealed that different parts of the body are under tighter genetic control than other parts; for example, Buckler and Green (2008) found that in a sample of 1,533 two- to nine-year-olds, height and weight were more influenced by the environment than head circumference.

THE HUMAN GROWTH PATTERN

Human skeletal growth occurs in a specific pattern that is probably best understood when compared to other **primates**. All primates grow slowly. Slow growth is likely linked to an expanded period of brain growth; that is, a large brain is energetically expensive, and so body growth is slow to allow for the expensive brain tissue to grow. Humans grow more slowly than all other primates, including chimpanzees (Walker et al. 2006; Zollikofer and Ponce de León 2010). Chimpanzees, for instance, often start to reproduce at around thirteen years of age, while hunter-gatherer females are capable of reproducing at around nineteen years of age (Walker et al. 2006). Although environments of abundant resources may speed up the process of growth in humans, all humans have a similar pattern of growth (Walker et al. 2006). The slow growth of humans, which is seen in late **molar** eruption and late **cessation** of bone fusion, can be called ontogenetic retardation, or **neoteny** (Zollikofer and Ponce de Leon 2010). The long growth period also allows for females to live many years prior to **menarche** (which is the onset of menstruation and ovulation) and obtain bone masses that are high enough to help prevent osteoporosis in later years because, after all, not only do humans take a long time to grow up, but they live long as well (Wang et al. 2005).

Although the overall trend of human growth can be summarized in one word—*slow*—human growth rates vary throughout life. Human skeletal growth decelerates during infancy compared to gestational growth and then maintains a steady rate throughout childhood until the adolescent growth spurt (Bogin 1999). Chimpanzees differ from this pattern; they have two periods of accelerated growth (whereas humans have just one growth spurt): one right after infancy and then a second one during their "childhood" (Bogin 1999).

LONG BONE GROWTH

Human skeletal growth either occurs at intramembranous ossification sites or at **endochondral** growth plate sites. Human skeletal growth at intramembranous ossification sites, such as the fontanels in the infant skull, occurs when bone is created in membranes. In contrast, longitudinal growth occurs at the growth plates by endochondral bone formation; basically, bones get longer because at the epiphyses there are growth plates that allow for formation of bone over a cartilaginous model. Figure 2.1 displays a tibia with unfused epiphyses. These thin layers of cartilage at the epiphyses account for nearly all bone growth in humans, with the exception of the flat bones like the skull. Endochondral growth plates consist of three zones: resting, **proliferative**, and hypertrophic. The resting zone contains hyaline cartilage; cell reproduction of **chondrocytes** (i.e., cartilage cells) occurs in the proliferative zone; and chondrocytes grow in the hypertrophic zone. Chondrocytes farther from the epiphyses stop dividing and become hypertrophic (Nilsson and Baron 2004). Hypertrophic chondrocytes are invaded by osteoblasts, which then remodel the cartilage into bone (Nilsson and Baron 2004). The growth plate is resorbed by sexual maturity; it appears that estrogen causes the fusion of the epiphyseal growth plate and the resorption (Weise et al. 2001). Throughout one's life, the rate of chondrocyte proliferation decreases, and it has been suggested that estrogen is a key hormone in this process (Nilsson and Baron 2004; Weise et al. 2001).

The reason for cessation of growth is unknown; some researchers have suggested that there is a finite amount of chondrocyte proliferation (Nilsson and Baron 2004). Fusion of the epiphyseal plates is not the cause of the cessation

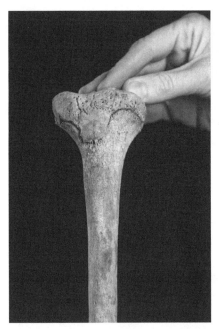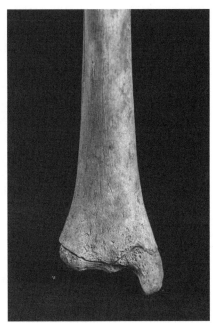

Figure 2.1. Unfused proximal (left) and distal (right) epiphyses on a tibia. The displayed tibia from a sixteen- to eighteen-year-old male has clear epiphyseal lines that demonstrate that the epiphysis had yet to fully fuse. Photographs by Daniel Salcedo.

of growth (Parfitt 2002). Although estrogen seems to have an effect on chondrocyte proliferation and epiphyseal fusion, these effects are distinct. Fusion occurs because growth has stopped, but growth has not stopped as a result of fusion (Parfitt 2002). Epiphyseal fusion is an active process that can occur only after growth ceases.

Cessation of growth and fusion of the epiphyseal growth plates vary by sex and race; for example, females stop growing earlier than males, and within males African-Americans and Mexican-Americans stop growing earlier than European-Americans (Crowder and Austin 2005). The reasons for the variation are likely part genetic and part environmental.

OSTEOLOGICAL INDICATORS OF GROWTH

The most common ways to assess deviations from normal growth in skeletal collections is to look at **Harris lines** and stature, but rickets and **anemia** are also considered important health factors that may impact growth. These indicators of growth can provide osteologists with information regarding increased and decreased health. When examining these traits, anthropologists look at trends over time in relationship to changes in subsistence patterns, increased urbanization, and differences within and between populations, especially when class, race, or sex differences may affect health. Although genes play the major role in growth, the environment has an influence on growth as well (Bogin et al. 2007). Furthermore, genetic studies on twins may overestimate the genetic influence on growth since most studied individuals are likely to live in an affluent Western society.

Environmental influences always deter individuals from reaching their genetic potentials; that is, environmental stressors, such as malnutrition, will result in shorter individuals than those of the same gene pool with no such stressors. The environment reduces the optimal outcome. This may seem confusing when talking about the increase in height over time, but some of the environmental stressors of the past that hindered optimal growth have declined, such as childhood diseases that can be prevented through vaccination. Although there are many causes of growth retardation or inhibition (such as Cushing syndrome, **celiac disease**, **thyroid** problems, **anorexia nervosa**, growth hormone deficiency, kidney problems, and intrauterine growth restriction), nutritional deficiencies and infections were likely the main reasons for retarded growth.

Malnutrition

In order for healthy growth to occur, bones' adequate energy requirements must be met. Energy is obtained through **carbohydrates** and **proteins**. Bone growth requires **amino acids**, which construct proteins and are most easily obtained through animal flesh consumption (Prentice et al. 2006). Minerals required for bone growth include calcium, phosphates, magnesium, and zinc (Umeta et al. 2003). Ions of copper, manganese, carbonate, and citrate are also required to maintain bone health (Prentice et al. 2006). **Vitamins** that are important for bone growth include vitamins C, D, and K; these vitamins are involved in crystallization of bone and collagen formation.

The three most common disorders of bone health include **stunting** (which will result in short stature), rickets (or the adult equivalent, **osteomalacia**), and osteoporosis with its related fractures (Prentice et al. 2006). Osteoporosis will be addressed in the next chapter since it is a problem found mainly in adults. Stature and Harris lines will be addressed as ways to assess stunting and cessation or slowing of growth. Stunting, which is defined as less than two standard deviations below the mean for growth, results in short adult stature, causes labor problems in the stunted individual due to the limited size of the uterus, and is linked to early death (Prentice et al. 2006). The small uterus can also lead to **trans-generational** short stature since the offspring will be born with low birth weight or prematurely (Prentice et al. 2006).

Stunting has been linked to a lack of zinc, protein, iron, and vitamin A. Umeta et al. (2003) found that stunting resulting from zinc deficiency is common in Africa; breastfeeding for longer than six months without sufficient supplementation may result in this deficiency. In developing countries, where breastfeeding is prolonged and animal flesh is scarce, zinc deficiency is common (Rivera et al. 2003). Vitamin A and iron seem to require severe deficiencies to affect growth, but only a mild deficiency in zinc causes stunting (Rivera et al. 2003; Umeta et al. 2003). In developing countries, like Tanzania, iron deficiency is rampant, and the cause is likely a diet consisting of too many grains and not enough meat (Mamiro et al. 2005). Truswell (1985) mentioned that protein and iron deficiencies can be cured by ferrous sulphate tablets; but these tablets are not well tolerated, and thus animal flesh consumption is the best source of these nutrients to prevent stunting. In the bioarchaeological record, iron deficiency that results in anemia is assessed by cribra orbitalia and **porotic hyperostosis**; these two traits occur on the skull. The appearance of both cribra orbitalia (which occurs in the eye sockets) and porotic hyperostosis (which occurs mainly on the parietal bones of the skull) is a sieve-like **porosity**. Figure 2.2 shows healed cribra orbitalia. These traits have been found in skeletons dating back three hundred thousand years and are common in many collections (Larsen 1987). Increases in cribra orbitalia and porotic hyperostosis in populations where maize agriculture was adopted fits in with the fact that maize inhibits iron absorption (Larsen 1987). Furthermore, populations that consumed high levels of beef and fish, such as prehistoric southeastern

Figure 2.2. Cribra orbitalia. This thirty-one- to forty-year-old male has evidence of childhood anemia in the form of healed cribra orbitalia. Photograph by Daniel Salcedo.

US populations, had fairly low levels of cribra orbitalia and porotic hyperostosis. Nonetheless, some anthropologists, such as Walker et al. (2009) and Rothschild (2012), have mentioned that iron deficiency is not the likely cause of cribra orbitalia and porotic hyperostosis; they argue that these traits relate to infections and **parasites** instead of diets.

Differentiating infections and parasite loads from dietary deficiencies is difficult. There is a feedback loop, too; for example, zinc deficiency that causes stunting also results in a decrease in immune system effectiveness and thus increases the chance of infectious diseases that can cause growth to cease (Caballero 2002). Solomons (2003) stated that the lack of nutrients in developing countries is more likely due to a lack of sanitation than a lack of appropriate foods and that treating the food availability without increasing sanitation will not help African children. Infectious diseases, including parasites, will be addressed in chapter 8.

Rickets, which is the result of vitamin D deficiency, can result in low bone mineral content and slow collagen formation. Diagnosis of rickets in skeletal samples includes porosity and deformation at the diaphyseal ends in long bones, short stature, and lower body deformities (Mays et al. 2009). Rickets, which occurs in children, is the failure of calcification of growth plates and causes deformation of the ends of diaphyses (Prentice et al. 2006). Although vitamin D is activated by sunlight, rickets is still a common disease in children even where sunlight is common. At present, usually the cause of rickets is cultural; veiling and a lack of outdoor activities has resulted in an increase in rickets. Munns et al. (2012), for example, discussed the increase in rickets in Australian immigrant children and noted that veiling was a factor in rickets risk. Children are also predisposed to rickets if they lack adequate calcium, phosphate, and **folate** (Prentice et al. 2006; Truswell 1985). In the bioarchaeological skeletal record, rickets is found in some medieval populations, especially in northern locations like the United Kingdom (Mays et al. 2009; Pearson 1997; Veselka et al. 2012). Adults seemed to have gotten more vitamin D than children due to foods, such as fish, eel, veal, and mushrooms, that were not consumed by children regularly (Pearson 1997). Additionally, it appears that during the medieval period in Europe, infants born in the winter were more likely to have rickets (Mays et al. 2009; Pearson 1997; Veselka et al. 2012).

Although nutritional deficiencies are common in developing countries, in developed countries doctors may fail to diagnose malnutrition because they are not familiar with the symptoms. Carvalho et al. (2001) provide two case studies of malnutrition in the United States from affluent families; the first case study involved a protein-deficient twenty-two-month-old child. The parents were well educated, and when their child developed **eczema** and started to vomit, they assumed it was a reaction to cow's milk. The mother thus replaced milk with rice milk. The child developed protein deficiency (known as kwashiorkor); during this time he had stunted growth. The other case came from a seventeen-month-old child who had rickets; the parents preferred soy milk to milk, and thus the child also received soy milk; his lumbar started to curve (called **kyphosis**), his bones started to become weak and osteoporotic, and his wrist bones flared. Both children were treated for their nutritional deficiencies; their growth resumed, and other symptoms dissipated.

Harris Lines

Harris lines, which are also known as growth recovery lines and growth arrest lines, were first described by Henry Albert Harris in 1931. He described these lines as dense lines that run parallel to the physis and whose appearance is a result of temporary growth cessation that is provoked by stresses (such as malnutrition) or medical conditions (such as diabetes) (Harris 1931). The lines develop before puberty; once growth is completed, growth cannot be interrupted. Although the lines are made prior to puberty, they may also persist well into the fifth decade of life (Hummert and van Gerven 1985). Conversely, bone remodeling can also erase the appearance of Harris lines (Hummert and van Gerven 1985; Laor and Jaramillo 2009). The formation of Harris lines can be understood in three basic steps. First, there is a decrease in chondrocyte proliferation in the proliferation zone of the epiphyseal plate; this decrease in chondroblasts is said to be triggered by stresses (Cunningham and Stephen 2010). Second, osteoblast activity continues, and thus the collagen fibers around the periosteum are subjected to tensile pressures that stimulate bone remodeling (Cunningham and Stephen 2010). Third, the slowing down of cartilage conversion to bone, but continuation of mineralization of developed trabecular bone, results in horizontal dislocation of trabecular bone on the metaphyseal side of a growth plate (Cunningham and Stephen 2010; Laor and Jaramillo 2009). When growth resumes, the line is pushed away from the growth plate. Osteologists can determine the age at which the Harris line was formed by its location on the bone (Garn and Schwager 1967; Laor and Jaramillo 2009). The number of Harris lines may also indicate whether multiple periods of stress occurred (Steffian and Simon 1994).

Identifying Harris lines can be performed using X-rays or magnetic resonance imaging (MRI) (e.g., Laor and Jaramillo 2009; Steffian and Simon 1994). The opaque horizontal lines are most frequently found on tibiae (shinbones) and femora (thighbones), but other sites such as arm bones, vertebrae, and iliac crests have also been studied (e.g., Blanco et al. 1974; Cunningham and Stephen 2010; Sajko et al. 2011). Correct identification and counting may involve determining whether the opaque line is complete or partial (Cunningham and Stephen 2010; Grolleau-Raoux et al. 1997) and excluding other line formations, such as **zebra lines** which involve more bones and are of shorter duration (Etxebarria-Foronda and Gorostiola-Vidaurrazaga 2013). Additionally, bone growth itself or excessive bone formation can also mimic Harris lines (Pförtner and Hövel 2003). It is important to use multiple bones and both left and right sides in determining Harris line frequencies since **asymmetry** is common, and thus by looking at just one side, osteologists may miss evidence of Harris lines (Hughes et al. 1996).

Harris line etiologies and correlations with other stress indicators are controversial. Mays (1995), for example, found no correlation with Harris lines and bone length in a medieval British sample. But some archaeological studies have found correlations with Harris lines and early deaths (e.g., Nowak and Piontek 2002). In some prehistoric populations, such as in Nubia between AD 550 and 1450, all children had Harris lines, but only about two-thirds of the adults did (Hummert and van Gerven 1985); this leads to the question of whether the adults' bones experienced bone remodeling to erase the lines or whether those who survived to adulthood experienced less stress as children. The high rates of Harris lines in various prehistoric (e.g., Nubia, AD 550–1450; Hummert and van Gerven 1985), proto-historic (e.g., Maori, AD 1600–1800; Chapple 2005), and historic samples (e.g., medieval Swiss; Papageorgopoulou et al. 2011) have led some anthropologists to conclude that Harris lines may be normal and are not indicators of stress at all. Furthermore, Alfonso-Durruty (2011) conducted an experimental test on rabbits to see if Harris lines could be produced; fifteen rabbits were fed normally, fifteen rabbits were underfed consistently, and fifteen rabbits were fed and then starved. Then all the rabbits were given ample food. The rabbits were fed the same type of food. There were no differences in Harris lines in the different rabbit groups, but it may be that the moderate undernutrition was not sufficient to cause Harris lines. A lack of correlation between stress indicators and Harris lines in some studies has been suggested to be the result of individual variability; a variety of factors such as when the stress occurs, nutrition, and childhood illnesses will determine who experiences growth cessation and who will have Harris lines (Hummert and van Gerven 1985).

Clinical data show clearer links between Harris lines and stresses. Hewitt et al. (1955) examined data from 650 children from the Oxford Child Survey and found that Harris lines correlated with illnesses. Blanco et al. (1974) examined 1,412 children between the ages of six and seven years from rural Guatemala and found that poor nutrition and high parasite loads corresponded to short statures and Harris line frequencies in X-rays of the radius. Males, interestingly, were more affected by stress than were females. Another clinical study found that ethanol alcohol consumption during growth years resulted in short stature and Harris lines (González-Reimers et al. 2007).

Although the bioarchaeological studies have mixed results on the effectiveness of Harris lines to indicate stress, many studies use Harris lines as evidence of nutritional stress. Steffian and Simon (1994) argued that first-millennium Alaskan foragers experienced recurrent nutritional stress; they found that in ninety-four bones examined, forty-three had Harris lines, and nearly all of these individuals had multiple lines. Chapple (2005) examined the Maori of New Zealand and found that between AD 1600 and 1800, Harris lines were more frequent in females than in males, which was attributed to sex differences in food access. Ameen et al. (2005) looked at both medieval and contemporary Swiss samples; in the medieval sample, 100% of the children had Harris lines, and in the contemporary sample, half of the children had Harris lines. The authors attributed the high frequency of Harris lines in the medieval sample to a lack of protein since other skeletal indicators of anemia were present. The contemporary children with Harris lines were either deficient in vitamin D (and had rickets) or were going through psychological stresses, such as parental separation.

In recent clinical studies, the causes of Harris lines seem to be separated by developing-world causes and developed-world causes. In India between 1963 and 2005, for example, Teotia and Teotia (2008) found Harris lines and rickets as a result of lack of vitamin D exposure. The authors argued that the lack of vitamin D was in part latitudinal (that is, the northern Indian samples had higher frequencies of rickets than the southern Indian samples) and in part cultural. The full covering with clothing of women and children can lead to rickets; further, cultures that do not allow these members of society to leave home also have increases in rickets and Harris lines in the homebound individuals. In addition, children who are institutionalized have high rates of growth arrest and rickets in India. Blanco et al.'s (1974) study on Guatemalan children found Harris lines and tied these growth arrests to low nutrition and high parasite loads.

In the United States and Europe, for the most part, studies on Harris lines have focused on fracture-related causes and medicinal-drug-related causes. Children who have a fractured bone, especially a leg bone close to the epiphyseal plate, may develop Harris lines (Ecklund and Jaramillo 2002; Hynes and O'Brien 1988). School sports and the increase of intensity in children's athletics may result in more fractures and thus more Harris lines. The Harris lines that are fracture related are local, but still clinicians worry about whether these bones are more prone to injuries and deformity as a result of weakened bone morphology (Hynes and O'Brien 1988).

Other studies have focused on the use of bisphosphonates and pamidronates. **Bisphosphonates** and **pamidronates** are drugs that help prevent bone loss and are commonly given to prevent osteoporosis, but they are also given for childhood congenital diseases that affect bone growth, such as **cerebral palsy** and **osteogenesis imperfecta**. Osteogenesis imperfecta is a genetic disease that results in a defect in collagen building; symptoms include bowed legs, bone fractures (especially prior to puberty), and curvature of the spine (Marini 2003). Both of these types of drugs have been found to increase bone growth and density, but both have also caused Harris lines (Etxebarria-Foronda and Gorostiola-Vidaurrazaga 2013; Rauch et al. 2004). Yet Etxebarria-Foronda and Gorostiola-Vidaurrazaga (2013) suggested that the lines from the drugs are not actually Harris lines, but more accurately described as zebra lines. They are not the result of temporary growth cessation and resumption, but rather they are the result of extra bone growth. Zebra lines tend to be more extensive than Harris lines, and they may disappear faster.

Stature

Some anthropologists have discussed whether the effects of reduced stature as a result of adverse environments may be an adaptive strategy to ensure survival and reproduction. Bogin et al. (2007) examined whether shortened legs (which are measured through the cormic index that takes into account overall stature and sitting height) are an adaptation to adverse environmental conditions (such as malnutrition, infections, economic oppression, and even high altitude) or whether shortened height is just a cost of living under stress. They found that stressful environments result in shortened legs but that this is not an adaptation. Shorter legs did not increase the survival or reproduction rates of these individuals.

Stature, or height, has been used by social scientists and medical practitioners to assess health. Stature is a complex trait that involves classic polygenic inheritance and environmental components. It appears that the highly heritable trait of stature involves at least forty-four loci; still, even though 80% of the variability in stature may be explained by genetics, there is still much to explore in stature with regard to health (Weedon and Frayling 2008). The environment's impact on stature seems to decrease the odds of reaching the ideal or genetic potential, and thus environmental stress results in short individuals. Stature decreases have been linked to malnutrition and diseases (Kemkes-Grottenthaler 2005). Additionally, taller individuals have a survival advantage. Using a sample of nearly three thousand European skeletons dating from AD 500 to 1900, Kemkes-Grottenthaler (2005) found that tall individuals led longer and healthier lives than short individuals. Furthermore, short individuals had higher risks of cardiovascular disease, **respiratory disease**, and **coronary heart disease**. However, Kemkes-Grottenthaler (2005) stated that the relationship between stature and health is not causal; tallness does not increase health, but rather short individuals experienced environmental assaults that prevented their growth. Interestingly, McIntyre (2011) reported that although shorter individuals are at greater risks from cardiovascular diseases and diabetes, taller individuals are at greater risks from cancers. The increased cancer risk may be in part the result of a longer life. These environmental assaults that result in short stature often start during the prenatal period and can even be trans-generational. For example, short females tend to have difficult childbirths and small offspring who grow up to be short adults (Kemkes-Grottenthaler 2005). Given the importance of a healthy environment and the environment's impact on stature, stature can be used to assess general health.

There are various ways to measure stature, and different parts of the body are more likely to be affected by the environment. Trunk length is least likely to be affected by environmental influences, and leg length is most likely to be affected by environmental influences (Kemkes-Grottenthaler 2005; McIntyre 2011). Thus, sitting height (also known as the cormic index) is commonly used to determine environmental impact on stature in living populations. The formula for the cormic index is (sitting height / stature) × 100. Additionally, distal elements are more sensitive to environmental pressures than are proximal elements; in other words, stresses will cause the tibia to stop growing before they cause the femur to stop growing. The crural index, which is the ratio of the length of the tibia to the length of the femur, can be used to assess tibial shortness. These measurements should be taken within populations and not compared between populations since evolutionary climatic pressures have resulted in different body proportions for different populations (see chapter 1). Within living populations, the measurements are fairly straightforward, and people can be measured in multiple ways quickly. In past populations, anthropologists can choose to run regression formulae by using a specific bone or set of bones and reconstructing the likely height of the individuals (e.g., Cardoso and Garcia 2009; Giannecchini and Moggi-Cecchi 2008; Trotter and Gleser 1951); using regression formulae is especially necessary when comparing past populations with living populations or data from skeletons to data from individuals who were living at the time of data collection. But regression formulae are tricky because they are based on formulae that were derived from individuals with known heights, and thus they are usually based on forensic collections, such as those mentioned in chapter 1. These populations are not representative of prehistoric populations and not even representative of many living peoples. Nevertheless, in order to determine secular changes in stature, regression formulae are our best bet. When comparing past populations with one another, simple measurements of long bones can suffice.

Stature only gives information about health and environmental stresses from conception to cessation of growth. Short stature indicates stress, but not what specifically caused the stress. Plus, catch-up growth may erase short stature and, thus, this indicator of childhood stress.

With the knowledge that stature is sensitive to the environment, osteologists can examine secular trends in stature to determine health issues of past and present peoples. Secular trends in stature have been of interest to anthropologists to examine the effects of climate, agriculture, urbanization, and disease. For example, Formicola and Holt (2007) looked at stature decreases in Europeans during the **Last Glacial Maximum** (around twenty thousand years ago). They tried to assess other health indicators to figure out whether the shorter stature in these remains compared to earlier remains and later remains was adaptive or an indicator of stress. They concluded that there appeared to be little evidence of stress (such as Harris lines) and thus deduced that shortness in these remains was an adaptive strategy. Conversely, Mummert et al. (2011) looked at the decrease in stature with the adoption of agriculture in the Western Hemisphere and found that it corresponded with an increase in population density, infectious diseases that spread in unsanitary locations, and malnutrition. Giannecchini and Moggi-Cecchi (2008) examined over a thousand skeletons dating from the ninth to the fifteenth century in Italy; they found that from the Iron Age to the Roman period there was a decrease in stature that was likely linked to early urbanization and a decrease in sanitation, whereas no change in stature resulted from the Roman period to the medieval period. Not all examinations of secular trends show a decrease in stature; Shin et al. (2012) found that Koreans dating from the fifteenth to the twentieth century showed no changes in stature. Pretty et al. (1998), who looked at skeletal remains of Australian Aborigines from 9,800 years ago to AD 1850 and Aborigine data on stature from AD 1996 and 1997, also found no significant differences in stature. Although Aborigine stature may have remained static, Australian children increased in height from 1899 to 1999; the increase in stature, however, slowed down between the 1950s and 1980s (Olds and Harten 2001). Cardoso and Garcia (2009) examined femora from early medieval period (fifth to ninth century AD) Portugal and found that children of the **Dark Ages** (another term for the medieval period) were not shorter than twentieth-century Portuguese children and that medieval adults were actually taller than the twentieth-century adults. It turns out the Dark Ages in Portugal were likely not as bad as previously imagined, whereas the 1970s in Portugal were probably worse than reported. One possible reason for the lack of an increase in stature in Portugal is that children of the twentieth and twenty-first centuries mature earlier than children one hundred years ago (Cardoso et al. 2010); using a sample of 521 present-day children and 114 children from a century ago, Cardoso et al. found that dental roots are developed 1.22 years earlier in boys and 1.47 years earlier in girls compared to the past population.

Short stature in the past is mainly linked to disease load and malnutrition, but currently short stature in developed countries may be the result of either over-abundance or scarcity in food, space, or sanitation.

In developing countries, populations are still facing problems of nutrition and sanitation that lead to stunting and short stature (Solomons 2003), and even in Russia the fall of communism has led to short stature in Siberian populations who are more isolated, not covered by the collective, and no longer receive food and other basic supplies through helicopter drops (Leonard et al. 2002). Yet, as early as 1941 anthropologist Howard V. Meredith noted an increase in stature in the United States; using data from individuals born between 1920 and 1927 compared to data from individuals born between 1930 and 1937, Meredith found that stature increased by three-quarters of an inch, and weight increased by three pounds. Meredith reported that the increase in stature in the United States over the last fifty years (starting in 1890) was likely a result of better child care and an increase in nutrition. Trotter and Gleser (1951) looked at the Terry Collection and military records to assess changes in stature in the United States; the Terry Collection birth dates ranged from 1840 to 1919, and the military records included birth dates from 1900 to 1924. Trotter and Gleser (1951) found that stature increased beginning in the twentieth century; the increases were in spurts, which may have been an artifact of the sample or linked to particular advances, such as vaccinations. From 1900 to after World War II, Kimura (1984) noted an increase in stature in Japanese that was mainly dependent on leg length. Kimura suggested this was a result of improved nutrition. However, the trend of increased stature in Japan seems to have halted; by 1963, sitting height stopped increasing, although leg length increased until the generation born between 1970 and 1980. Meadows Jantz and Jantz (1999), who examined stature from the Huntington Collection, the Terry Collection, World War II records, and the Forensic Anthropological Database, found that both males and females had a drop in height in the mid-nineteenth century followed by an increase in height from the early twentieth century to the 1970s. Lower limbs, which are a better proxy for stature than upper limbs, showed the most increase, and white males gained the most leg length over time. Meadows Jantz and Jantz (1999) suggested that males are more sensitive to environmental changes than females. Hoppe et al. (2006) suggested that a large degree of the increase in average stature in industrialized countries is a result of the easily accessible nutrients, such as protein, in cow's milk. Most stature changes from the previous six to seven decades (from 1900 to about 1960) were positive, but in the last fifty years the trend has reversed.

Multiple studies have reported a decrease in height in the last fifty or sixty years in the developed countries using both large samples of thousands of individuals, such as National Health and Nutrition Examination Surveys

(NHANES) in the United States or data from schools and the National Health Services Central Register in the United Kingdom, and smaller samples from clinical populations. Organizations such as the WHO also conduct large surveys on BMI, puberty age, and stature. Ahmed et al. (2009) examined European data from the late nineteenth century to the present and found that although stature increased initially due to better health, nutrition, and sanitation, in the last forty years there has been a reversal, with shorter stature. Ahmed et al. (2009) attribute this decrease in stature to an increase in obesity. In Germany, Finland, England, Scotland, and Austria there has been a 4% increase each decade in obesity among children. A study on births between 1963 and 1970 in the United States found that obesity is linked to short adult stature (Sunder 2008), and Cardoso and Padez (2008) found that in Portugal an increase in weight has led to a decrease in height. Part of the reason for obesity's link to short stature is that obese children's epiphyses fuse earlier. Shalitin and Phillip (2003) found that obesity is linked to fast growth; thus, children are tall for their age, but upon adulthood they are actually shorter. And Komlos and Breitfelder (2008) noted that in the United States, blacks mature faster and have higher BMIs than whites; blacks also tend to be tall for their age, but shorter as adults than whites.

The link between obesity and early epiphyseal fusion may lie in the link between fat and puberty (Carel et al. 2004). Obesity coupled with early puberty and early puberty coupled with low birth weight have been linked to a decrease in stature in many developed countries (e.g., Denmark, Aksglaede et al. 2009; Portugal, Cardoso and Padez 2008; Spain, Ibáñez et al. 2006; United Kingdom, Sandhu et al. 2006; United States, Komlos and Breitfelder 2007). Obesity has been linked to greater adipose tissue, which increases the likelihood of early puberty. Leptins, which are found in **adipocytes** (fat cells), signal puberty development.

At what age early puberty (or precocious puberty) occurs depends on who is asked. Historical research has given us later ages of puberty than current Western populations. Research on Dutch populations has found that during the **Middle Ages** puberty occurred at fourteen years of age, but in the nineteenth century puberty occurred at around seventeen years of age; the onset of puberty is affected by the environment (Keizer-Schrama and Mul 2001). Gluckman and Hanson (2006) suggested that about half of the variance in the onset of menarche is determined by genes, but that leaves much room for environmental influence. Gluckman and Hanson estimated that in the Paleolithic period menarche occurred between seven and thirteen years of age, whereas during the Neolithic period menarche probably occurred slightly later, around nine to fourteen years of age. Their calculations are based on limb lengths, life span, and comparative primate data. Recently, age at puberty has decreased; in the United States, for example, breast development occurred at around eleven years of age in 1969, whereas it occurs in eight year olds now. Kaplowitz and Oberfield (1999) have argued for a change in the age definition of early puberty. Early puberty as defined by Marshall and Tanner in 1969 was before eight years of age in females and before nine years of age in males; Kaplowitz and Oberfield would like the white female age to be dropped to seven years of age and the black female age dropped to six years of age (and no change for males). They noticed that in a sample of seventeen thousand children, puberty occurred between 8.5 to 13 years in 95% of the sample; thus, 8 years is normal since it is so shortly before 8.5 years of age. Kaplowitz and Oberfield (1999) also noted that early puberty may be longer in duration, and when this occurs only minimal height differences are present. Komlos and Breitfelder (2007) disagreed; they hypothesized the shorter stature in US adults compared to Dutch adults is linked to the later puberty in the Dutch sample. They suggested that just because puberty is occurring earlier in many developed countries, that does not mean it is ideal or normal; the Centers for Disease Control and Prevention (CDC) and the WHO purport that best practices in lifestyle should govern definitions of precocious puberty rather than defining normal by frequency of occurrence (Komlos and Brietfelder 2007).

Early puberty is associated with short stature in many developed countries. Styne (2004) noted that certain populations in the United States, such as African-Americans and Hispanics, have higher BMIs and reach puberty earlier than European-Americans. Scientists have found that early puberty, especially in low-birth-weight children, leads to short stature. Ibáñez et al. (2006) found that children in northern Spain who had low birth weight were more likely to reach puberty early, and their puberty did not last long; thus, they reached their adult stature early and were shorter than their cohorts. McIntyre (2011) reported on differences between developing countries and the United States; he found that in developing countries earlier puberty was not linked to shorter stature, but their earlier puberty was not as early as in the United States. In developed countries, early puberty and menarche were linked to shorter legs and, especially, shorter tibiae. Sandhu et al. (2006) looked at a sample of over a thousand individuals born between 1927 and 1956 in the United Kingdom, and they reported that higher BMI was linked to earlier puberty and associated with decreased leg length and height. In short, fat increases the chance of early puberty, which in turn increases the chance of early epiphyseal closure and shortened stature. As a contrast to the early puberty trend, elite gymnasts, especially females, have delayed puberty and shortened stature as a result of excessive training and low body mass (Erlandson et al. 2008).

Administering drugs, such as gonadotropin-releasing hormone (GnRH) agonists, to stop early puberty has successfully led to increased stature (e.g.,

Carel et al. 2004; Palmert et al. 1999). Clinical trial researchers, such as Palmert et al. (1999), have argued for use of GnRH agonists to slow down puberty because of the link between early puberty and obesity; still, even these clinicians accept that the children had high BMIs prior to the onset of their puberty.

Some people have argued that the West's decrease in the age of puberty is not a result of diet or BMI, but rather a result of chemicals and the foods consumed. Endocrine disrupters are either plastics (such as BPA or PPA), plants (such as soy), or growth hormones given to animals (rBST) that mimic estrogen. Environmentalists have been concerned that these chemicals and hormones, which are used globally, may be affecting children's health. The evidence for puberty being triggered by plastics is inconsistent; Lomenick et al. (2010) found no evidence of early puberty with an increase in the endocrine disrupters in urine, but Wolff et al. (2010) found a weak correlation in a large sample of 1,151, with early breast development and pubic hair and three types of endocrine disrupter chemicals. Den Hond and Schoeters (2006) found mixed results; in Michigan exposure led to a decreased age of menarche, but not an earlier age of breast development; whereas in Italy there was no change in age of puberty onset. Collier (2000) reported that rBST fed to cows to increase their milk production is not passed to consumers of the milk, especially since pasteurization destroys the hormone. Findings on growth and soy show no consistent changes in puberty (Cederroth et al. 2010), although minor changes in males' testes and breasts have been noted (Tan et al. 2006).

Catch-Up Growth

Catch-up growth is rapid growth that occurs after a deceleration or cessation of growth due to adversity in the environment (Silventoinen et al. 2011; Williams 1981). The growth that occurs during catch-up growth is at a velocity above the statistical limits of normal growth for the specific age, and it is of limited duration (Williams 1981). Catch-up growth occurs in children up to the time when their epiphyseal plates fuse; catch-up growth can only occur after a period of growth retardation when the reason for the retarded growth has been removed (Williams 1981). Catch-up growth may allow individuals to reach their genetic potential even when adversity exists. If the individual is returned to the original pre-retardation growth curve and reaches the original optimal height, then catch-up growth is said to be complete (Williams 1981). Most of the time, catch-up growth still results in individuals who are shorter than their optimal stature; thus it is incomplete catch-up growth (Williams 1981). Complete catch-up growth is more likely if the stress that invoked retardation occurred early in life (Adair 1999). Additionally, more severe and longer-term stress reduces the odds of individuals reaching their optimal growth (Williams 1981). But even when catch-up growth occurs early, not all individuals reach their optimal stature; Knops et al. (2005) found that in a Dutch sample of 753, stunted growth occurred even when there was catch-up growth, and that infants who were small for their gestational age were more adversely affected than pre-term infants who had appropriate weights for their gestational age.

Although osteologists have determined that during catch-up growth there is an increase in the proliferation of chondrocytes at the growth plates, the mechanics of catch-up growth remain poorly understood (Nilsson and Baron 2004). Gafni and Baron (2000) proposed that catch-up growth occurs when glucocorticoid hormones slow linear growth by suppressing chondrocyte proliferation and delaying epiphyseal fusion. They hypothesized that glucocorticoid hormones put growth on hold for conservation of nutrients for vital functions and that when growth inhibition is stopped, growth is resumed by the release of glucose and **insulin**.

In skeletal samples, catch-up growth has been implicated in the erasure of short stature, and thus anthropologists looking at skeletal remains have suggested that other indicators of stress, such as **vertebral canal** size and **enamel hypoplasia**, are better traits to examine to determine stress. For example, Watts (2011) examined the longevity, vertebral canal size, and stature of skeletal remains from the United Kingdom's Fishergate House, which was an area that was occupied by individuals of low socioeconomic status, during a time of urban expansion (tenth to fifteenth centuries) and found that vertebral canal size and longevity were negatively correlated, but stature did not correlate with the other health indicators. Vertebral canal size is determined by nine years of age and thus can be used to determine childhood stresses; a small canal is an indication of retarded growth. Watts (2011) suggested that the poor health indicators were a result of childhood malnutrition and infections, whereas the lack of a short stature is indicative of catch-up growth. Thus, stature would not have been a good health indicator to use with the Fishergate sample. In Asian skeletal samples, similar results have implied that catch-up growth has erased the evidence of stress with regard to stature. Temple (2008) looked at the health status of prehistoric Japanese populations dating from 5,000 and 2,300 years before present. Jomon remains from both the western and eastern areas of Japan were examined for enamel hypoplasia (a stress indicator similar to Harris lines but which is prevalent on the teeth) and stature. The western Jomon had scarce resources and relied mainly on seasonal plants for food, with maritime food supplementing their diet, whereas eastern Jomon consumed terrestrial meats, plants, and maritime foods (and had an abundance of food). Not surprisingly, the western Jomon had a greater frequency of enamel hypoplasia compared to the eastern populations, but interestingly the western and eastern populations had no difference in stature. Temple attributed this lack of stature difference to catch-up growth. Domett and Tayles (2006) similarly looked at enamel hypoplasia and stature in 4,000-year-old Bronze Age Thai skeletal remains compared to 1,800-year-old Iron Age Thai skeletal remains. Both populations had enamel hypoplasia, but the Iron Age males were taller than the Bronze Age males, which the authors attributed to catch-up growth during the Iron Age sample.

Catch-up growth has long been assumed to be a positive factor in human development. Anthropologists have assumed that it is a sign that the individuals who were under stress recovered; they got over their infections or they received more nutritious foods. Clinical studies, such as Graham and Adrianzen (1972) and Whitehead (1977), have emphasized the importance of catch-up growth in children from developing countries. Graham and Adrianzen (1972) looked at children in Peru who were severely malnourished and found that with intervention the children were able to improve their height even if intervention was late; at eighty-eight months (seven years and four months), children still experienced catch-up growth. Whitehead (1977) emphasized the importance of enough protein to allow catch-up growth. Recently, however, catch-up growth has been seen as a negative factor for long-term health.

Multiple studies from Europe, the Americas, Asia, and Australia have confirmed that catch-up growth correlates with **type II diabetes**, obesity, and cardiovascular disease. Most of these studies have focused on low-birth-weight infants whose gains have been a result of medical intervention to promote catch-up growth. For example, Ong et al. (2000) looked at British individuals from the Avon Longitudinal Study of Pregnancy and Childhood and found that in a sample of fourteen thousand, catch-up growth in the first two years of life was linked to higher BMI and larger waist circumferences later in life. Eriksson et al. (1999) found that for children born between 1924 and 1933, low birth weight followed by catch-up growth by age seven resulted in an increase in coronary heart disease. Low-birth-weight babies lack muscle and put on more fat, which may be part of the cause of the rise in type II diabetes and heart disease in low-birth-weight offspring who are able to achieve catch-up growth.

Cettour-Rose et al. (2005) proposed that the reason for the negative effects of catch-up growth is because infants who experience catch-up growth become hyperinsulinemic during the rapid growth. Hyperinsulinemic means there is too much insulin in the blood; insulin is a hormone produced in the pancreas that regulates sugar and converts sugar into energy. Thus, too much insulin then converts glucose, which is the sugar from foods, to fat for conservation of energy for periods of famine. When no famine arises, the fat remains. Hyperinsulinemia causes type II diabetes because the body becomes insulin resistant when the body does not utilize the insulin well; thus, glucose builds up in the blood rather than being converted to energy. Dulloo (2008) suggested that in the past, catch-up growth and the conservation of glucose would have actually been adaptive because famines were likely, but in the current developed nations, this is no longer the case.

Not all studies, however, find that catch-up growth is negative; Victora et al. (2001) looked at a large sample of 3,582 Brazilians from birth to 3.5 years of age and found that catch-up growth resulted in lower child mortality and a decrease in hospital stays. They suggested that in developing countries it is important to promote catch-up growth through medical and non-profit organization intervention. In the Philippines with a sample of over two thousand children, Adair (1999) found, conversely, that without intervention a moderate level of catch-up growth still occurs. However, it may be that the catch-up growth resulted from changes in the environment, such as the increase in sanitation that occurred in the Philippines during the study period (Adair 1999).

CONCLUSIONS

Compared to earlier populations during prehistory and medieval periods, environmental changes have drastically improved the chances of reaching optimal growth. The control of infectious diseases, the increase in sanitation, and the availability of food resources have resulted in healthy growth and tall adults. The introduction of agriculture likely led to the first decrease in optimal growth, especially since foods such as rice and maize are low in protein (Widdowson et al. 1991); a second decrease in growth likely occurred with urbanization, which led to higher population densities and a decrease in sanitation. Furthermore, urbanization led to vitamin D deficiency that resulted in rickets. Yet an increase in stature and optimal growth reversed the negative trend; starting after the 1940s child growth became uninterrupted and sped up; this was especially true for developed countries (Widdowson et al. 1991). Optimal growth still eludes the developing countries, and sub-Saharan Africa leads the world in nutritional deficiencies and infectious diseases that affect children (Solomons 2003). Conversely, developed countries have reduced their optimal growth due to an excess of calories; the excess has led to early puberty that results in shortened stature as a result of early fusion of growth plates. Obese children become short-legged adults; this may reverse the gains in stature from earlier times. Finally, low-birth-weight and premature infants, which are more likely to survive now, have also resulted in a new wave of catch-up-growth children. Catch-up growth has been linked to later health problems, such as type II diabetes and cardiovascular diseases.

CHAPTER 3

Adult Bone Health

IN THIS CHAPTER, BONE MINERAL DENSITY (ABBREVIATED AS BMD) WILL BE discussed. A similar measure to bone mineral density—bone mass—can be used both in growth studies and in studies of adults that look for healthy bone indicators (Heaney 2003). Although both bone mass and bone mineral density can be affected by childhood health, this measure is mainly important for assessing adult morbidity. Bone mineral density is the amount of mineral in bone per square centimeter and should not be used in growth studies (Heaney 2003). The mineral content in bone ensures that bone is strong enough to avoid fracturing under normal conditions. The three forms of low bone mineral density that will be covered in this chapter are osteomalacia, osteopenia, and osteoporosis.

OSTEOMALACIA

Osteomalacia is a bone disease in which there is an excess of unmineralized bone (also known as osteoids) (Chalmers et al. 1967). Osteomalacia differs from osteoporosis and osteopenia because it involves defective bone formation rather than just an imbalance of bone deposition and resorption. In osteomalacia, calcium is not absorbed, and so the osteoids never become mineralized; bones remain soft.

Osteomalacia is difficult to differentiate from other bone diseases in skeletal collections (Brickley et al. 2007; Foldes et al. 1995). Clinical diagnoses of osteomalacia involve skeletal pain, muscle weakness, low bone mineral density, and Looser's zones (Chalmers et al. 1967; Gifre et al. 2011). **Looser's zones** are made of horizontal translucent bands and **pseudofractures** that are several millimeters thick and tend to be present on both left and right sides (Brickley et al. 2005, 2007; Chalmers et al. 1967). Additionally, when fractures are slow to heal, osteomalacia may be the culprit (Chalmers et al. 1967). In osteomalacia, bone mineral decreases first in the peripheral skeleton and then in the spine. The ribs and pelvis are also affected in osteomalacia (Brickley et al. 2005; Chalmers et al. 1967; Rosin 1970).

The most common cause of osteomalacia is vitamin D deficiency (Chalmers et al. 1967; Gannagé-Yared et al. 2000; Rosin 1970). The lack of vitamin D prevents calcium absorption; vitamin D enables the body to absorb dietary calcium and utilize it for bone hardening (Chalmers et al. 1967). In the bio-archaeological record there are few cases of osteomalacia; some of the cases come from cold, cloudy, northern locations, such as Poland (e.g., Haduch et al. 2009), whereas others come from urban medieval and post-medieval Europe, especially the United Kingdom (e.g., Brickley et al. 2007; Pinhasi et al. 2006). However, other cases occur in very sunny climates; for example, Foldes et al. (1995) reported on a Bedouin osteomalacia case dated at around AD 500 in the Negev desert of Israel. They suggested the severe lack of mineralized bone is a result of a lack of vitamin D due to the culture of veiling. Brickley et al. (2007) suggested that osteomalacia may have been more frequent in the past, but the lack of bone mineral density would cause the bones not to preserve; this taphonomic issue cannot be resolved, and thus bioarchaeologists may always be missing cases of osteomalacia. Yet Robins (2009) noted that even in cloudy Europe an individual with dark skin only needs one to three hours three times a week of sun exposure to activate vitamin D sufficiently; this amount of sun exposure was likely easily achieved by hunter-gatherers, farmers, and many other workers. With light skin, three times a week of ten to fifteen minutes of sun exposure would be sufficient to fend off vitamin D deficiency (Robins 2009).

In present populations, vitamin D deficiency is more common in Europe than in the United States; this difference is likely a result of the climatic differences and the fact that fortification of foods in the United States is more prevalent than in Europe (Gannagé-Yared et al. 2000). The most common reasons for osteomalacia in present populations are cultural; for example, Asian and Middle Eastern females who wear veils are more likely to have osteomalacia than non-veiled females (Gannage-Yared et al. 2000; Jokar et al. 2008). Sometimes vitamin D deficiency in Asians is exacerbated by a **vegetarian** diet (Dandona et al. 1985), a low-fat diet (Rosin 1970), or a medical condition (Jabbar et al. 2009). Beyond cultural causes of osteomalacia, it can also be caused by intestinal malabsorption diseases, such as celiac disease (Gifre et al. 2011), and by kidney problems (Kanis 1981). Tumors can also lead to osteomalacia; some rare tumors cause hormones to inhibit kidney functions (Lewiecki et al. 2008). Kidneys are used in the process of turning inactive vitamin D to active vitamin D, which is known as calcitrol (Kanis 1981; Lewiecki et al. 2008).

Medical intervention for diseases, such as epilepsy, can lead to osteomalacia. Macallan et al. (1992) described a female patient with osteomalacia who took anti-convulsant drugs to stop her epileptic seizures; the patient was also East Indian, veiled, and a vegetarian. While she took vitamin D supplements, her bone formation did not suffer from osteomalacia, but upon stopping her supplements, the symptoms of osteomalacia returned. Recent increases in gastric bypass surgery may also increase the cases of osteomalacia (e.g., Basha et al. 2000). Although high BMI seems to protect bones from losing bone mineral, gastric bypass surgery may increase osteomalacia by increasing malabsorption of dietary sources of vitamin D, such as fish. Previous research has shown that gastric surgery increases osteomalacia risk (Ravn et al. 1999).

OSTEOPOROSIS AND OSTEOPENIA

Osteoporosis and **osteopenia** involve bone weakening due to an imbalance of bone deposition and bone resorption. The resorption of bone that has been previously constructed outpaces the deposition of new bone in osteopenia and osteoporosis. Additionally, osteoporosis and osteopenia result in changes in the micro-architecture of trabecular bone; in trabecular bone, with osteopenia and osteoporosis there is a loss of connectivity that results in intracortical cavities (Seeman 2002). These intracortical cavities put the hip, vertebrae, and forearm at risk of fractures. Osteopenia and osteoporosis can be thought of as being on a continuum. The medical definition of osteopenia and osteoporosis is related to a comparative bone mineral density value. Bone mineral densities, which are represented by standardized scores, are used to determine osteopenia and osteoporosis. Osteoporosis occurs when the change in bone mineral density is significant enough that the individual in question is at risk of fractures (Brunader and Shelton 2002; Mølgaard et al. 1997; Seeman 2002).

Osteopenia and osteoporosis are silent diseases until a fracture occurs. Fractures mainly occur in the vertebrae, the proximal femur, and the wrist. Vertebral collapse, especially at the lumbar vertebrae, are the most commonly found osteoporotic fractures in the bioarchaeological record, but in the clinical environment many vertebral fractures go undiagnosed (Mays 2006a). Figures 3.1 and 3.2 show two types of vertebral collapse. **Colles' fractures**, which occur at the lower forearm near the wrist, and femoral neck fractures are more frequently the first indicators of osteoporosis in clinical settings. Figure 3.3 shows a healed Colles' fracture. The most disabling of osteoporotic fractures are femoral fractures; bioarchaeologists do not often find femoral fractures, but Curate et al. (2011) found evidence of these fractures in prehistoric Portugal. Furthermore, looking at more recent samples, such as the Hamann-Todd Collection and the Coimbra Collection, hip fracture rates of the nineteenth and twentieth centuries seem comparable to modern epidemiological rates.

Figure 3.1. Thoracic vertebral collapse. The top two vertebrae show decreased vertebral body heights (compared to the third vertebral body) that are characteristic of vertebral collapses. These vertebrae come from a forty-plus-year-old female who may have had post-menopausal osteoporosis. Photograph by Daniel Salcedo.

Measuring Bone Mineral Density

Clinical methods to measure bone mineral density involve using radiographs or sonograms (Brunader and Shelton 2002). Dual energy X-ray absorptiometry (also known as DEXA or DXA) is the most accurate way to measure bone mineral density; these X-ray scans are usually completed on the lumbar, the femoral neck, or the radius. DEXA measures trabecular bone, which is more likely to fracture (Brunader and Shelton 2002). A slightly less precise scan than the DEXA is the DPA (dual photon absorptiometry). Quantitative computed tomography (qCT), which is an X-ray taken at 360 degrees, is another option for measuring bone mineral density; usually, the lumbar is measured (Brunader and Shelton 2002). The benefit of qCT is that it can give information about both trabecular and cortical bone. However, qCT exposes the patient to greater radiation levels than DEXA or DPA. Ultrasonography uses sound rather than radiation to take a picture of bone; the architecture of the trabecular bone at the heel bone (i.e., the **calcaneus**) is assessed when

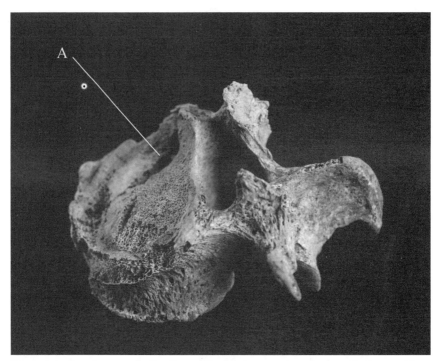

Figure 3.2. Lumbar vertebral collapse. The lumbar vertebra displayed is from a thirty-five- to forty-four-year-old male who could have experienced vertebral collapse, which is evident from the downward sloping of the vertebral body, from bone loss or from a hernia (A). Photograph by Daniel Salcedo.

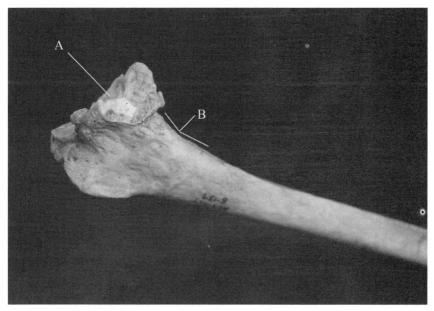

Figure 3.3. Colles' fracture. This distal radius from a forty-one- to fifty-year-old female has secondary arthritis (A), which is likely the result of a healed Colles' fracture that can be seen in the slightly misaligned distal end of the shaft (B). Photograph by Daniel Salcedo.

employing ultrasonography. Ultrasonography does not expose the patient to radiation, but the method is not universally accepted and standards of comparison are not available (Brunader and Shelton 2002).

The bones that are most commonly examined to assess low mineral density in the clinical literature are the femoral neck, the lumbar vertebrae, and the forearm; these bones are most at risk of fracture as well (Brunader and Shelton 2002; Seeman 2002). In order to evaluate risk of fractures, the bone mineral densities are converted to standardized scores (either z-scores or t-scores), which are based on numbers below the standard deviations for a healthy thirty-five-year-old female (Brunader and Shelton 2002). According to the WHO, normal bone mineral density value is within one standard deviation of the mean, osteopenia is a bone mineral density value between 1 standard deviation and 2.5 standard deviations below the reference mean, and osteoporosis requires a bone mineral density that is more than 2.5 standard deviations below the reference mean.

Bioarchaeologists have been assessing bone mineral density to determine changes in bone health throughout time and locations. Investigations of bone mineral density on skeletal remains often utilize techniques that are used in clinical settings; DEXA and DPA scans of lumbar vertebrae, femoral necks, and radii are common methods employed to scan bones of past peoples (e.g., Foldes et al. 1995; Holck 2007; Zaki et al. 2009). Yet some researchers have used methods not commonly used in medicine; for example, X-rays of metacarpals and femora (e.g., Beauchesne and Agarwal 2011; Lees et al. 1993; Mays 2006a) have been employed. Bones have also been sliced and then X-rayed (e.g., Agarwal et al. 2004; White and Armelagos 1997). These alternative methods may have been done because the more advanced radiographs were not available, but the difference in methods may make comparisons with living populations difficult.

Types of Osteoporosis

Osteoporosis can be divided into secondary, which is caused by medical conditions (such as multiple sclerosis) or treatments (such as use of lithium) that interfere with the attainment of **peak bone mass**, and primary (i.e., idiopathic juvenile, idiopathic young adult, and involutional adult osteoporosis), which is when bone loss is either genetic or age related. Involutional (or age-related) osteoporosis is then divided into type I and type II; type I is when there are trabecular bone changes in **post-menopausal** females, and type II is when there are changes in both the cortical bone and the trabecular bone. Type I occurs in women between forty-five to seventy years of age, whereas the more severe type II osteoporosis occurs in women and men over seventy years of age (Zaki et al. 2009).

HORMONES AND OSTEOPOROSIS. Involutional osteoporosis is mainly the result of fluctuating hormones through life. As a result, males and females have different patterns of osteoporosis. Females are more likely to get osteoporosis than males, and males are afflicted with osteoporosis later in life. Both non-sex and sex hormones play a role in bone health throughout humans' lives. Hormones that regulate calcium homeostasis are called calciotropic hormones; the primary ones are vitamin D, parathyroid, and calciton (Madimenos 2011). Calciton, which is produced in the thyroid, removes bone through osteoclastic functions. Growth hormones, which stimulate cell production, and insulin-like growth factor, which stimulates osteoblasts, both play a role in deposition of new bone (Madimenos 2011). Insulin-like growth factor is also the primary hormone that determines sexual dimorphism upon puberty. After puberty has occurred, female bone mineral density is mainly regulated by estrogen, whereas male bone mineral density is controlled by both estrogen and later androgen (Callewaert et al. 2010). These hormonal differences result in female bone deposition occurring on the endosteum whereas male bone deposition occurs at the periosteum; this radial expansion of male bones results in stronger and larger skeletons in males than in females (Callewaert et al. 2010).

For females, estrogen and progesterone fluctuate through life as a result of the menstrual cycle, pregnancy, **lactation**, and finally menopause. Estrogen increases intestinal absorption and retention of calcium, and it decreases bone resorption; a decline in estrogen levels increases porosity and bone loss (Madimenos 2011). Progesterone increases bone mineral density through osteoblast activity. Estrogen, progesterone, and calciotropic hormones fluctuate as a result of females' reproductive cycles. At menarche, the release of estrogen through the ovaries increases osteoblastic activity. Females who attain menarche early tend to have high BMD values (Madimenos 2011). Progesterone inhibits contractions during pregnancy, which results in an accrual of bone. Lactation alters the calciotropic hormones and lowers estrogen, which lowers BMD temporarily. However, the increase in prolactin and oxytocin while breastfeeding increases calcium absorption (Madimenos 2011). Finally, at menopause, estrogen is no longer secreted by the ovaries, and thus osteoclastic activities are not countered, which results in higher bone porosity and overall bone loss (Callewaert et al. 2010).

Although hormones play the primary role in osteopenia and osteoporosis etiology, there are many confounding factors that affect risk of bone loss. Factors such as peak bone mass, age, ethnicity, reproductive decisions, body weight, lifestyle, and medical practices all help to determine who is at risk for osteopenia and osteoporosis.

Ethnicity and Genetics of Bone Mineral Density

Studies on skeletal differences in bone mineral density and osteoporosis rates have resulted in some interesting questions regarding genetic and lifestyle influences on bones. For example, Mazess and Mather (1974) examined Eskimo bone mineral density and found it low in comparison to Europeans; the authors suggested that this low bone mineral density in Eskimos is the result of a high-protein diet and not a lack of vitamin D or calcium. And Jaleel et al. (2010) examined East Indian females and found that they have lower bone mineral density than their Caucasian counterparts, which was attributed to a low-calcium diet, vitamin D deficiency, **parity**, and genetics. They mentioned that East Indians are smaller and have lower peak bone mass than Caucasians. When examining black and white differences, multiple studies have found that blacks have higher bone mineral densities and fewer fractures than do whites (Mensforth and Latimer 1989; Nelson et al. 2000). These differences seem to be a result of genetics; black populations have higher calcium and potassium content in their bones and greater muscle mass. Another genetic population difference was found in a comparison between Han Chinese and Caucasians. In a sample of 1,131, Dvornyk et al. (2003) found that the Chinese population examined had **alleles** (which are alternative forms of the same gene) that resulted in lower bone mineral density than the alleles in the Caucasian population examined. Dvornyk et al. (2003) hypothesized that alleles coding for lower bone mineral density are not selected against since osteoporosis occurs late in life and thus does not affect reproductive success. The genetics of bone mass variation can be better understood through twin studies since populations differ in many ways, such as diet and activity level.

Studies examining differences between monozygotic (identical) and **dizygotic** (non-identical) twins can reveal the genetic influence of traits. Monozygotic twins are genetic clones, and thus, if a trait is completely determined by genes, then monozygotic twins should always share the same trait, whereas dizygotic twins would share the trait only half the time (like ordinary siblings). Using thirty-eight monozygotic twin pairs and twenty-seven dizygotic twin pairs ranging from twenty-four to seventy-five years of age, Pocock et al. (1987) determined that bone mineral density is highly heritable. Furthermore, different anatomical structures varied in regard to the genetic influence of bone mineral density; on the low end, forearm bone mineral density variance is 58% environmentally determined and 42% genetically determined, but on the high end, lumbar bone mineral density variance is 92% genetically determined. Other twin studies (e.g., Christian et al. 1989) support the high heritability of bone mineral density and peak bone mass.

Age and Peak Bone Mass

Osteopenia and osteoporosis risk increase with age, but the correlation does not arise until later in life (Krølner and Nielsen 1982). Pre-menopausal females, for example, show no correlation with bone mineral content and age in the lumbar region, but after menopause bone mineral content does correlate with age (Krølner and Nielsen 1982; Mazess and Barden 1991). When one ages, bone remodeling experiences imbalances for multiple reasons; this imbalance can express itself as an excess in bone resorption or a lack of bone deposition (Matkovic et al. 1994; Stini 1990). In type I osteoporosis, which occurs in females after menopause, the decrease in estrogen results in an increase in resorption that is most pronounced in trabecular bones (Stini 1990). Type II osteoporosis, on the other hand, affects both sexes and results in a decrease in **subperiosteal** bone deposition (Stini 1990). Usually type II osteoporosis only occurs in the seventh or eighth decade of life. The reason for the decrease in bone deposition may be a result of decreased calcium absorption (Stini 1990); however, another reason for osteoporosis in old age is the decrease in muscle strength, which is both controlled by **androgenic** hormones and calcium (Burr 1997; Stini 1990). Menopause, as well, results in a loss of strength and is coupled with the onset of type I osteoporosis (Burr 1997). It appears muscle loss may actually precede bone loss, and **longitudinal studies** suggest that muscle loss starts in the thirties while bone loss starts a decade later (Burr 1997).

Although osteoporosis occurs late in life, peak bone mass plays a vital role in determining fracture risk later in life. Mazess and Barden (1991) estimated that two-thirds of the osteoporotic risk in females can be explained by pre-menopausal bone health, especially peak bone mass; the other third of the osteoporotic risk is controlled by post-menopausal factors. Since bone is dynamic, the amount of bone mass varies throughout life; when an individual has reached the most bone mass that they will ever obtain, they have reached their peak bone mass. Peak bone mass is influenced by genetics and lifestyle, including diet and exercise. Although some researchers have theorized that peak bone mass can be obtained as late as thirty-something in women and in the forties in men (see sources in Stini 1990), other researchers have claimed that peak bone mass is obtained in the late teens. Matkovic et al. (1994) examined a sample of 265 pre-menopausal Caucasian females between the ages of eight and fifty years old and found that peak bone mass was obtained at around eighteen years of age. Møl-gaard et al. (1997) found in their sample of children from Copenhagen that bone mineral content increased until 17.4 years for males and 15.7 years for females. Overall, it appears that peak bone mass is positively affected by moderate physical activity; for example, Fehily et al. (1992) found that Yugoslavian females had higher peak bone mass values when they started sports before menarche.

In short, peak bone mass, which is obtained at a young age, is one of the main predictors of osteoporosis in old age.

Past populations should not have high rates of bone loss, due to their short lives. Yet some past populations had alarmingly high rates of osteopenia and osteoporosis. In the archaeological samples, it appears that early-onset (starting as early as the thirties for some samples) osteopenia and osteoporosis may be the result of low peak bone mass (Beauchesne and Agarwal 2011; Mays 2006a; Mays et al. 2006; White and Armelagos 1997). The lower peak bone mass in past populations, however, is likely not the result of a lack of activity. Thus bioarchaeologists have suggested that the lower peak bone mass is related to poor nutrition rather than low activity levels. Ericksen (1982), examining remains from the Terry Collection, found that, compared to the archaeological literature, twentieth-century individuals lost bone later than archaeological samples. Interestingly, even though bone loss occurs earlier in some prehistoric and medieval samples, the pattern of bone loss is essentially the same; for example, females lose bone earlier than do males (Beauchesne and Agarwal 2011; Mays et al. 1998).

There has been a recent increase in osteopenia and osteoporosis in Western countries. The increase can be in part explained by aging. Life expectancy has increased from sixty-five years in 1950 to seventy-seven years in 2010. According to the US Census between 2000 and 2010, the population sixty-five years and over increased at a faster rate than the total US population. This increase in age and osteoporosis has led to a proliferation of osteoporosis drug treatments. Many osteoporosis drugs are bisphosphonates and increase bone mineral density, especially vertebral bones, but in some individuals they can cause fractures in the metacarpals and femur (below the femoral neck) (Harris et al. 1999; Sutton et al. 2012). Other factors, such as lifestyle changes, are to be blamed for the increase in osteoporosis as well, such as a decrease in activity and an increase in smoking (Poulsen et al. 2001). Changes in parity may account for some of the changes in osteoporosis as well; after all, birthrates have been falling in Europe, parts of Asia, Canada, and the United States. Global fertility rates have decreased by 60% over the last five decades (Fineberg and Hunter 2013).

Pregnancy, Lactation, and Birth Control

Pregnancy and lactation require calcium for the growth and mineralization of the fetal skeleton and the production of breast milk. During pregnancy, the mineralization of the fetal skeleton needs about thirty grams of calcium (Kojima et al. 2002). Lactation uses even more calcium and does so for a longer time (Kojima et al. 2002). Lactation at low body weights usually triggers **amenorrhea**; amenorrhea as a result of low estrogen levels has been hypothesized to further harm bone mineral density. Estrogen is essential in reducing bone resorption and increasing bone deposition (Wiklund et al. 2012). However, some clinical studies have shown that calcium absorption increases during pregnancy (Cross et al. 1995) and that bone mineral density recovers from the stresses of pregnancy and lactation (Wiklund et al. 2012).

In the past, most females likely had multiple offspring. In the bioarchaeological record, Poulsen et al. (2001), examining Danish remains from AD 1000 to 1250, found that bone mineral density in females was low in comparison to modern Danes; they suggested the lower bone mineral density in these early Danish remains may relate to multiple pregnancies and lactation. And Mays et al. (2006) examined medieval English villages dating between the tenth and sixteenth centuries and found that female bone mineral density declined earlier than expected; this was also attributed to multiple offspring. Beauchesne and Agarwal (2011) examined a sample from Imperial Rome and found that females had lower bone mass than found in modern samples; the Roman females also lost cortical bone earlier than would be expected. The authors have suggested that the high parity (between five and eight children) of the females coupled with long lactation put these females at risk for osteoporosis. Additionally, they suggested that peak bone mass may have been low due to under-nutrition. This low peak bone mass and under-nutrition may be a factor in many early populations.

In clinical studies on parity and lactation, the results are varied. In some studies there appear to be no long-term effects; other studies show parity and lactation to be protective of bone health, and yet others show negative trends of bone mineral density associated with pregnancy and breastfeeding. Kojima et al. (2002) examined a large sample of over a thousand Japanese females who were both pre-menopausal and post-menopausal and looked for correlations with bone mineral density and reproductive history. They found parity had no effect on bone mineral density, but lactation did negatively impact bone mineral density in forty- to forty-four-year-olds. The lack of a correlation between bone mineral density and lactation in females over forty-four years of age suggests that females recover from the lactation-related loss of bone mineral density. In studies of populations who have large families, positive effects of parity and lactation have been reported. Streeten et al. (2005) used a sample of 424 Old Order Amish and found that parity correlated positively with hip bone mineral density and a decrease in fractures. Some of the results were likely the result of increased BMI, but in older females between ages fifty and fifty-nine, the parity and bone mineral density correlation remained even when controlling for BMI. Streeten et al. (2005) suggested that multiparity improves bone health by increasing weight, increasing calcium absorption, increasing cumulative estrogen exposure, and delaying menopause. Specker and Binkley (2005) looking at Hutterites found greater bone area with parity too; they suggested that the radius and femoral neck are more robust because of the increase in physical activity involved in carrying the baby during pregnancy and then carrying the infant after pregnancy. One may theorize that the women in the above studies are not representative of the larger populations; the Amish and Hutterites are unlikely to take medicines (including birth control pills), they rarely smoke or drink alcohol, and their main beverage is milk, which is a good source of calcium (Specker and Binkley 2005; Streeten et al. 2005). In Northern European countries, there have been studies on parity and lactation using the general population; Michaëlsson et al. (2001) looked at a large sample of Swedish females and found that parity decreased fractures in elderly females, but if the women had previously taken birth control pills, then fractures were actually higher with more offspring. And Wiklund et al. (2012) found that longer lactation in Finnish females was correlated positively with bone mass two decades after the child was born.

Western countries may not be ideal comparisons for past populations; thus the lack of a positive correlation with bone mineral density in bioarchaeology may be more similar to developing countries. Looking at a large Moroccan sample of 730 females, parity was associated with lower bone mineral density, especially in the spine (Allali et al. 2007). The Moroccan females had low levels of calcium intake, which may be similar to past populations. Yet the authors also mentioned that cultural factors may have increased osteoporosis; low bone mineral density was associated with veiling, a lack of sun exposure, and low levels of physical activity. Few studies on non-Western or non-Islamic populations exist, but Madimenos (2011) published BMD studies on the Shuar forager-horticulturalists of Amazonian Ecuador. The Shuar sample studied had no history of contraceptive use, they have a healthy diet, and they have much exposure to sunlight. Madimenos found that parity, lactation, and birth interval did not negatively affect the Shuar females. Any decreases in BMD seemed to be transient for the Shuar females, which corroborates the evidence from Western countries that suggests high parity and lactation are not hazardous to bone health.

In many modern societies, birth control pills or oral contraceptives have become a common part of many women's lives. Nearly a third of all childbearing-age women in the United States use oral contraceptives. Understanding the impact of birth control pills on women's bones is essential to understanding the osteoporosis risk on these women decades later. Hormonal birth control use during reproductive years has been linked to a higher bone mass density in post-reproductive years in some studies and the opposite in other studies. Wei et al. (2011) looked at a sample of 687 females who took progesterone and estrogen birth control pills, progesterone-only pills, and no pills. Using ultrasonography of the calcaneus, they found that using the combination progesterone and estrogen pill increased bone mineral density. Wei et al. did not find the increase in bone mineral density in progesterone-only pill use. Their results remained robust even when controlling for BMI, age, activity, and parity. Kritz-Silverstein and Barrett-Connor (1993) looked at a Caucasian sample of 239 females and found that the lumbar, hip, and radial bone mineral density as measured with DEXA and DPA were higher in females who took the pill six years or more compared to non-pill users. The theory behind the improved bone mineral density, according to Kritz-Silverstein and Barrett-Connor (1993), is that estrogen works to increase calcium absorption and decrease calcium loss. Scholes et al. (2011) examined the duration and dose of birth control pill users in a large sample of over 600 females between the ages of fourteen and thirty years old to determine the effects of birth control pills on bone health. They found that teen birth control users who were between fourteen and eighteen years of age did not differ from non-users in spinal or whole-body bone mineral density, but birth control users between the ages of nineteen and thirty years old had lower spinal and whole-body bone mineral density than non–birth control users. The results were especially prominent in low-dose oral contraceptives. They hypothesized that individuals who take the pill for longer periods of time and with a lower dose experience suppression of the estrogen peak that is essential in increasing calcium absorption.

In short, it appears that in the bioarchaeological sample, pregnancy and lactation led to lowered bone health in females, which is similar to what is found in developing countries. These negative correlations with reproduction and bone health are likely a result of under-nutrition and other environmental factors, whereas in developed countries, especially in populations that eschew medicines, alcohol, and smoking, bearing many children and long periods of breastfeeding improve bone health.

Health Foods and Bad Habits

In bioarchaeological studies, poor nutrition has been cited as one of the reasons for low bone mineral density. For example, Beauchesne and Agarwal (2011) argued the low peak bone mass in Imperial Romans was likely due to under-nutrition coupled with having many offspring and breastfeeding for long periods of time. White and Armelagos (1997) found that Nubian females with osteopenia dating between AD 350 and 550 had a different diet than females without osteopenia. And Mays (2006a) associated poor nutrition with low peak bone mass and an increase in osteoporosis in British females dating between the third and fourth centuries AD. However, not all bioarchaeological studies find a link between diet and bone mineral density or bone mass variation; when investigating osteoporosis in Native American remains of hunter-gatherer and agricultural populations of the Southwest, Perzigian (1973) reported that diet was not predictive of osteoporosis.

In the clinical literature, a diet rich with calcium, protein, phosphorous, manganese, zinc, and vitamins A, C, and D is associated with healthy bone mineral content. Welten et al. (1995) conducted a meta-analysis and found that overall calcium intake was positively associated with bone mass. And New et al. (2000) found fruit and vegetable consumption during childhood along with calcium intake increased bone mineral density. Although people in developing countries still face under-nutrition and do not get enough vitamins, minerals, or proteins (e.g., Allali et al. 2007; Jaleel et al. 2010; Olivieri et al. 2008), most people in developed countries do not have these same problems. Yet lifestyle decisions based on religion, health consciousness, or ethics have led some people to become **vegans** and vegetarians; these decisions may cause nutritional deficiencies that will adversely affect bone health. Although Rizzoli and Bonjour (2004) and Freudenheim et al. (1986) noted that there is a strong association between bone mass and protein consumption, some researchers (e.g., Hunt et al. 1989; Marsh et al. 1988) have found no negative effects on bone health as a result of vegetarianism. For example, Marsh et al. (1988) found that at age eighty, Seventh-day Adventists who are vegetarian tend to have a higher bone mineral density than **omnivores** of the same age. Their advantage may lie in the high calcium intake; many Seventh-Day Adventists consume milk with each meal. Furthermore, they tend not to smoke, drink alcohol, or drink caffeine. Craig (2009) hypothesized that vegetarianism and veganism are not as bad for bones as previously assumed because individuals who eschew meat tend to consume high levels of potassium and magnesium, which inhibits bone resorption. Additionally, there appears to be a link between vitamin K, which binds calcium to bone matrix, and prevention of hip fractures; vitamin K is easily obtained through leafy greens. Yet, other research has called into question the validity of these diets as being sufficient for bone health. Fontana et al. (2005) found that raw-food vegans had lower bone mineral density at all sites than omnivores. And Promislow et al. (2002) found that there was a positive association with meat protein and bone mineral density and a negative association with vegetable protein and bone mineral density. Chiu et al. (1997), looking at Taiwanese nuns, found that they had lower spinal bone mineral density than omnivores of the same ethnicity and age. Finally, a meta-analysis by Ho-Pham et al. (2009) found that vegetarians had about 4% lower bone mineral density than omnivores, which puts vegetarians at greater risk of fractures.

Research on alcohol, smoking, and caffeine in past populations has been nearly non-existent; sometimes bioarchaeologists have mentioned that the lack of these vices may explain higher bone mineral densities in prehistoric or historic populations compared to modern populations. For example, Agarwal et al. (2004) examined a British medieval sample and suggested that their relatively high bone mineral density was in part due to a lifestyle of activity without steroids or tobacco. Also, modern alcohol may be much stronger than the beers and ales consumed in medieval Britain. Conversely, Holck (2007) found that medieval Norwegians, but not prehistoric or Viking period Norwegians, had greater bone mineral density than modern Norwegians; this result surprised Holck because he thought smoking and alcohol consumption might have a negative effect on modern bone density and thus expected higher bone mineral densities in the earlier populations.

Experimental research on bone health has found that smoking, drinking alcohol, and drinking or consuming caffeine has negative effects on bone mineral density and bone mass. Research on rats by Broulik et al. (2010) demonstrated that alcohol can cause osteoporosis; alcohol consumption decreased bone mineral density and changed the architecture of trabecular bone in their rat population. Alcohol's effects on bone seem to be the greatest when an individual starts drinking prior to achieving peak bone mass, but even drinking heavily as an adult increases the risk of low bone mineral density and fractures. For example, Kanis et al. (2005) found in a large study of over sixteen thousand males and females that alcohol consumption increases osteoporosis risk, but the relationship is not linear. Moderate drinkers who had two drinks or less a day had no negative effect on their bones, but heavy drinkers were at greater risk of low bone mineral density and fractures. Interestingly, some researchers have reported that moderate drinking in elderly females is actually beneficial to bone health (e.g., Turner 2000).

Smoking seems to incur only negative effects on bone health. The more a person smokes and the earlier smoking is initiated, the greater the risk of fractures and osteoporosis (Hollenbach et al. 1993; Taes et al. 2010). A meta-analysis by Law and Hackshaw (1997) found no difference between pre-menopausal smokers and non-smokers, but after menopause the negative effects of smoking were evident. In post-menopausal women, smoking increases the lifetime risk of hip fracture by nearly double compared to non-smokers; they estimated that one in eight hip fractures can be attributed to smoking. Smoking is not just bad for females; males who smoke are also at greater risk for fractures than non-smokers (Taes et al. 2010). Smoking alters trabecular bones' micro-architecture, thins cortical bone, and reduces bone mineral density (Hollenbach et al. 1993). Smoking seems to increase the stress hormone cortisol, which lowers bone mineral density. Also, smokers seem to have hampered calcium absorption.

Coffee and cola are two of the most consumed beverages in the world. The effect of caffeine on bones is not resolved. Some studies have reported a negative correlation with caffeine and bone health. Harris and Dawson-Hughes (1994) used DEXA to analyze bone mineral density in a sample of 205 females to examine the effects of caffeine; they found that females who drank more caffeine had greater bone loss. Interestingly, there was a positive correlation between caffeine consumption and calcium consumption as a result of dairy added to the caffeinated drink. For females who consumed adequate calcium, there was no bone loss, but for those who had high caffeine consumption coupled with low calcium consumption, bone loss was accelerated. Rapuri et al. (2001) looked at a sample of nearly five hundred elderly females over three years and found that high caffeine consumption led to an increase in bone loss at the spine. Furthermore, they discovered that individuals with an allele that interrupts vitamin D production had greater bone loss with caffeine consumption than those without this allele. Nevertheless, many studies have not found the caffeine–bone loss relationship. It appears that a lack of calcium consumption, rather than a loss of calcium in the urine caused by caffeinated drinks, leads to bone loss (Heaney 2002). Lloyd et al. (1997) noted that it is difficult to control for confounding factors in determining whether caffeine consumption harms bone health; yet they chose a sample of women without a history of smoking, alcohol drinking, or hormone use and found that even women who drank eight glasses of caffeine a day did not have lowered bone mineral density.

Physical Activity and Body Mass Index

Physical activity seems to increase bone mineral density (Bennell et al. 1997; Morris et al. 1997). Muscle usage helps to initiate bone remodeling, which results in thicker diaphyses and increased trabecular density (Ruff et al. 2006; Woo et al. 1981). Bone remodeling theory posits that when muscle insertion sites are subjected to stress, blood flow is increased, which stimulates bone-forming cells. From the clinical and biomechanical literature, osteologists have gleaned that activity through muscle use is an essential component in maintaining bone strength because muscle usage places stress on bones that is necessary to activate osteoblasts (Burr 1997; Ruff et al. 2006; Woo et al. 1981). Therefore, the reason bone mineral density is associated with activity lies with bone remodeling concepts, such as Wolff's law, that assume when a muscle insertion or origin site is subjected to stress the result is an increase in blood flow and a consequential stimulation of bone-forming cells.

In the bioarchaeological literature, Zaki et al. (2009) found that working-class females in ancient Egypt had better bone mineral density scores than females who were in the higher classes; they suggested it was because the working females engaged in more activity than did the high-class females. Interestingly, male workers were worse off than high official males; perhaps they were over-worked. Agarwal et al. (2004) examined agricultural British male and female lumbar vertebrae during the medieval period and found that the medieval sample had fewer fractures than in modern British populations, which the authors suggested related to higher levels of activity as a result of agricultural subsistence. Lees et al. (1993) examined femora from British females dating from AD 1792 to 1852 and compared the results with a modern British sample; they found that the eighteenth- and nineteenth-century females had better bone health than did the modern comparative sample. Lees et al. attributed this difference to higher levels of activity in the past population. Yet not all research shows this trend; as mentioned before, some bioarchaeological research has found that past peoples had lower bone mineral density, which is usually associated with poor nutrition coupled with parity.

In the clinical research, activity has been linked to an increase in both peak bone mass and bone mineral density later in life. Most studies emphasize the importance of exercise during childhood years to obtain a healthy peak bone mass that will prevent osteopenia and osteoporosis in later years, but Chow et al. (1987) provided evidence that even post-menopausal females can increase their bone health with aerobics and strength training. Even in research that has not found an increase in bone mineral density, a decrease in fractures has been linked to exercise started as late as the seventh decade of life (Korpelainen et al. 2010). Thus, it appears exercise—in the past and present—is useful for obtaining and maintaining healthy bones, but too much exercise can be harmful.

Young female athletes sometimes experience a triad of deleterious effects from over-exercising. This triad consists of too little energy availability, low bone mineral density, and menstrual dysfunction. Miller et al. (2012) estimated that between 1% and 5% of female athletes experience the triad. Female athletes tend to reach menarche later, have a lower bone mineral density, and show an increase in low-impact fractures that are not related to high-risk activities (Ackerman et al. 2011; Miller et al. 2012). Amenorrhea and irregular menstruation as a result of low estrogen levels are associated with low bone mineral density (Ackerman et al. 2011; Miller et al. 2012). Female athletes who have extremely low body fat are subject to having low levels of leptin hormones, which then prevents puberty since estrogen is partially produced in adipocytes. Although most female athletes (and their coaches) know of the connections between body fat, amenorrhea, and bone health, they are not likely to change their habits or gain fat if they are still interested in being competitive (Miller et al. 2012).

Although school athletics have become more competitive and have grown in the recent past, lack of activity and obesity in particular are much bigger problems than low body weight and amenorrhea. According to the WHO

(2013), the obesity rate has nearly tripled from 13% in 1960 to 34% in 2007. The effects of body weight and in particular high body mass indices on bone health are counterintuitive. Although high lean body mass is correlated with high bone mineral density scores (Morris et al. 1997), obese children have higher bone mineral contents and bone mineral densities than non-obese children (Leonard et al. 2004). Ravn et al. (1999) noted that thinness and low bone mass are correlated, especially in post-menopausal females. Additionally, Thomas and Burguera (2002) found that a higher BMI in post-menopausal women was associated with higher bone mineral density. Rosen and Bouxsein (2006) also reported on bone mass correlations with weight and BMI, which they think may be linked to mechanical loads on bones carrying excess weight. Bone mass, however, is negatively affected by obesity when it is coupled with metabolic diseases that are common in the obese, like type II diabetes (Pollock et al. 2011). Adults who undergo gastric bypass surgery experience a decrease in bone mineral density; doctors usually tell patients to increase their calcium and vitamin D consumption post-surgery, but this does not seem to be sufficient to stave off the bone loss (Coates et al. 2004).

CONCLUSIONS

Although much of bone mineral content and density is determined by genes, lifestyle impacts bone health as well. In past populations, poor nutrition coupled with having many children and breastfeeding for long periods of time seemed to decrease bone health and cause early-onset osteoporosis. When nutrition was good, past peoples had better bone health than many modern populations, which is likely due to the higher physical activity levels coupled with sunlight exposure, especially in non-urban settings, in the past populations. In present-day samples, bone health can be harmed in multiple ways; osteomalacia as a result of vitamin D deficiency is seen in cultures where females are covered up and stay in the home. Osteopenia and osteoporosis, on the other hand, seem to increase due to smoking, lack of activity, and fad diets. Having children (when proper nutrition is available), it turns out, seems to protect bones by increasing calcium absorption. The recent increase in BMI has led to higher levels of bone mineral density, but only when it is not coupled with diseases. And when doctors and patients think medical intervention can help us out, they find that gastric bypass surgery actually promotes bone loss, and drugs to prevent osteoporotic vertebral fractures can increase femoral shaft fractures. Finally, regardless of lifestyle decisions and genes, bone loss is inevitable when humans get older; as populations continue to age, higher osteoporosis rates can be expected.

CHAPTER 4

Childhood Injuries

In the developed world, injury has surpassed infections as the leading cause of death in children; in developing countries, infections are still a leading cause of death, and the rates of injuries are often unknown due to religious practices such as faith healing (Glencross and Stuart-Macadam 2000). In the developed nations, between 7% and 27% of children have experienced injuries that affect the skeleton; the most common of these injuries are fractures (Glencross and Stuart-Macadam 2000). The CDC also listed over-exertion, especially in children between ten and fourteen years of age, as a frequent cause of non-fatal injuries (Borse and Sleet 2009). In this chapter, I will address how four major issues (falls, abuse, sports, and obesity) affect childhood injury patterns.

DETECTING CHILDHOOD TRAUMA

Clinicians use magnetic resonance imaging (MRI) and radiographs to diagnose childhood injuries; MRIs allow doctors to accurately assess the extent of the injury and what soft tissues are involved in the injury (Balassy and Hörmann 2008). X-rays and CT scans are more effective when examining bone fractures only.

The most common fracture sites in children are lower limbs and distal radii (Balassy and Hörmann 2008). When a child experiences an injury, there is more variation in how bone can be affected than when an adult is injured; for example, the injury can result in deformation or breakage, whereas in adults deformation is rare (Balassy and Hörmann 2008; Meling et al. 2013). Children's skeletons are softer and more elastic than adults' skeletons; soft bones may experience bowing, buckling, and **greenstick fractures** (Meling et al. 2013). As illustrated in figure 4.1, in a greenstick fracture there will be bending and cracking rather than a complete clean break. The non-calcified growth plates can be separated from the diaphyses; if the growth plate is affected, then growth can be altered, and this may result in a shortened limb (Balassy

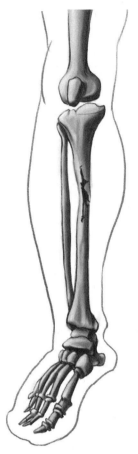

Figure 4.1. Greenstick fracture. A greenstick fracture rarely breaks the bone in two; it more often results in uneven fractures and splintering. Illustration by Vanessa Corrales.

and Hörmann 2008; Meling et al. 2013). About 15% of fractures in children involve the epiphyseal growth plates, and in 1% to 2% of these fractures, irreversible damage occurs due to angular deformation or shortening of the limb (Balassy and Hörmann 2008). Another common injury that may occur is an **avulsion**, which is when a muscle or tendon is pulled from the bone. Muscle and tendon pulls are common in **subadults** because youths exert strength and energy through muscle use, but the attachment sites are not ready for the forces (Balassy and Hörmann 2008; Donnelley et al. 1999). The most common location for avulsions is the lower limb, especially the knee region. **Osgood-Schlatter** trauma describes an avulsion at the **tibial tuberosity**, and Siding-Larsen trauma describes an avulsion at the patella (kneecap). Distal femur avulsion usually relates to over-exertion of the quadriceps tendon. Avulsions can lead to **periostitis**, which is a localized infection of the periosteum, and **heterotopic** ossificans, which is as an ossification of the tendon (Donnelley et al. 1999). Figure 4.2 shows a heterotopic ossificans.

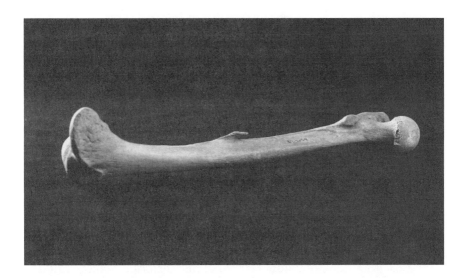

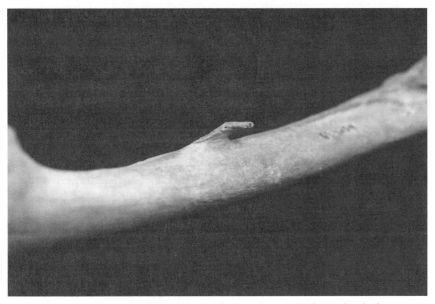

Figure 4.2. Heterotopic ossificans. The ossified tendon on the femur (top) of a twenty-one- to thirty-year-old female is a good example of an avulsion. Close up, one can see that it stems from the linea aspera (bottom). Photographs by Daniel Salcedo.

Bioarchaeological Methods of Detecting Childhood Trauma

Childhood injuries have always occurred, but childhood injuries have changed over time. Many causes of trauma, such as motor vehicle accidents and gunshots, are absent in past populations. Past children were unlikely to experience high-energy trauma (e.g., motor vehicle accidents), but they did experience low-energy trauma (e.g., falls, abuse, and stress). Furthermore, children in the past may have been treated as adults (Geber 2012; Perry 2005), and thus thirteen-year-olds may have experienced different stresses than present-day teenagers. However, even currently in developing nations children are often expected to engage in "adult" activities.

Research on childhood trauma in past populations is extremely scarce; bioarchaeologists concentrate on adult remains. The reasons for the neglect of childhood trauma in archaeological studies are many. First, children have smaller bones that are less likely to preserve than adult bones. Thus, children and especially infant remains are fewer than adult remains. Second, the **plasticity** of children's bones increases the chance of angular deformation even after death; thus, when **postmortem** fractures occur, they can resemble **antemortem** fractures (Glencross and Stuart-Macadam 2000; Walker 1983). In adults, when a bone is broken after death, the bone is brittle and dry, and fractures appear as irregular lines with no adhering bone flakes. Adult postmortem fractures can also be transverse and splintered. If an injury occurs before death in an adult, the bone retains some elasticity, and fractures tend to be spiral or running along the long end of the diaphysis. Antemortem fractures in adults also tend to have sharp, clean edges. On the skull, depressions are more common in antemortem fractures, whereas postmortem cases usually involve shattered bone. In adults, bones can retain some elasticity for several years after death in certain depositional environments, but this is rare. In children, however, bones are much more elastic; thus, even after death the bone is not as brittle, and so antemortem and postmortem fractures look alike. Third, bone remodeling and healing complicate the identification of childhood trauma in the archaeological record. Due to the great amount of variation and rapid bone remodeling of young individuals, it may be difficult for osteologists to identify trauma in young individuals. Children's bones heal more rapidly; their bones are more vascular and have more osteogenic activity. Healing time is strongly correlated with age, and some of the most common fractures, such as physeal fractures, are the fastest to heal (Glencross and Stuart-Macadam 2000). If a child has two or more years of growth left, then bone remodeling, healing, and growth will likely erase the evidence of trauma. Even when evidence of a fracture is visible, determining the age at which the fracture occurred is difficult because of the variation in growth due to environmental factors (Walker 1983). Finally, determining what is trauma and what is not can be difficult even in adults, but in children it is doubly difficult. A bone that has asymmetry in length or angular deformation can be mistaken for a congenital disease. If the bone was not completely broken and no deformation occurred, then no trace of the injury may be left when the individual is an adult (Glencross and Stuart-Macadam 2000). Epiphyseal injuries often result in cartilage trauma that leads to arthritis; one may mistake the arthritis for age-related osteoarthritis.

A few anthropologists have tried to determine causes of trauma in subadults and answer questions such as whether child abuse occurred in prehistory. For the most part, childhood trauma in the bioarchaeological record is reported as case studies (e.g., Blondiaux et al. 2002; DiGangi et al. 2010; Kacki et al. 2013; Wheeler et al. 2013). For example, in their research, Glencross and Stuart-Macadam (2000, 2001) utilized X-ray technology and clinical comparisons to attempt to diagnose humeral **condyle** fractures in seventeenth-century Ontario remains. They noted that using the same measurements as clinicians they were able to diagnose three rather than one humeral fracture that likely resulted from falls during childhood. Furthermore, they used patterns of clinical data to determine whether injuries were likely the result of childhood trauma in the seventeenth-century Ontario remains; for example, they commented that frequency, location, and nature of the fracture are influenced by age. There are a few exceptions to case studies. Jiménez-Brobeil et al. (2007) looked at the influence of terrain on fractures and found that there were more fractures in some—but not all—rough terrain Iberian prehistoric populations than in other flatland populations. Accurately determining the causes of these traumas in the bioarchaeological record requires comparative clinical data. It is likely that the bioarchaeological evidence for childhood trauma provides an erroneously low estimate of childhood injuries in past populations (Walker 1983).

PATTERNS OF CHILDHOOD TRAUMA: FALLS, ABUSE, SPORTS

Falls

Falls are the most common cause of non-fatal, low-energy trauma in children nowadays. It may be that presently there are numerous "falls" that are actually abuse and are reported by the abusive parents as falls. Although according to the CDC (2013) and the WHO (2013), motor vehicle transportation accidents are responsible for the most fatal injuries in individuals under nineteen years of age, when examining 9.2 million non-fatal emergency room visits, falls were the number one cause of injury (Borse and Sleet 2009). In most developed countries, falls may be less serious than other injuries, but in developing countries, falls from high places are not uncommon. For example, Gulati et al. (2012) reported that in Delhi, India, falls were the leading cause of death from unintentional injury, and the very young and very old were the most frequent victims. Gulati et al. (2012) stated that falls result from sleeping on roofs, flying kites on roofs, and climbing trees.

There are several cases of broken bones from falls in the archaeological record. In a second case presented by Glencross and Stuart-Macadam (2000), they noted a fracture of the humeral supracondyle that likely occurred as a result of a fall; this type of fracture is one of the most common found in the clinical record. Upex and Knüsel (2009) also found a single individual with multiple injuries; the medieval British skeleton had a broken rib and a greenstick-fractured femur, which they attributed to a fall. From a medieval graveyard in France, Kacki et al. (2013) found a 6- to 7.5-year-old with a right humeral fracture that affected the diaphyses. The shortened humeral neck caused by the growth plate injury may have been related to a fall; congenital deformity was ruled out since only one bone was affected.

Abuse

According to the WHO (2013), between 25% and 50% of the world's children experience physical abuse; rates of abuse, however, are difficult to assess due in part to cultural variation in what is defined as abuse. Males are more frequently abused than are females. Children under the age of four and adolescents experience higher rates of abuse than other age groups. Infants and toddlers are especially at risk since they cannot explain the causes of their injuries (Kemp et al. 2008). Unwanted children and children with special physical needs are also prone to abuse.

In the archaeological record, child abuse is rarely detected, which is surprising since historical records often support that child abuse was rampant in the past (Walker 1983). Wheeler et al. (2013) and Walker (1983) both point out that an absence of evidence for child abuse in past populations does not mean that child abuse did not occur, yet anthropologists have made that assumption. For example, Lovejoy and Heiple (1981) found that injuries were most common in individuals between ten and twenty-five years of age in a hunter-gatherer Native American population from Ohio; they further noted that there were no injuries in individuals under five years of age, which they suggested was indicative of a lack of child abuse. And Jiménez-Brobeil et al. (2007) reported a lack of intentional violence on children and suggested this may be an indicator of a lack of child abuse. Other anthropologists have found evidence for violence against children. Gaither (2012) found that children in Peru were subject to violence in both pre-European-contact and post-European-contact eras, but violence on children increased post-contact. Gaither suggested that the trauma experienced by children may have been the result of child abuse since stressed communities in modern populations seem to engage in higher rates of child abuse than non-stressed communities. Case studies have been used to support the long history of child abuse; Wheeler et al. (2013) published the oldest known case of child abuse in a first- to fourth-century skeleton in Egypt. Blondiaux et al. (2002) reported a case of child abuse in fourth-century Normandy, and during the late Roman period, a skeleton with multiple fractures provides evidence of possible child abuse (Lewis 2010). Most of the archaeological studies do not take into account the patterns of abuse found in current populations to reconstruct abuse frequencies in the past.

Only a third of abused children experience skeletal fractures, which means that archaeologists would be unable to identify two-thirds of the abused children as being abused. Walker (1983) thus remarked that in order to figure out rates of child abuse in prehistory, an extremely large sample is necessary. Later, however, Walker et al. (1997) suggested that excessive Harris lines and stunted growth may be an indicator of child abuse even if one does not find evidence of fractures. They also examined the skeletons of children who died from abuse and found that direct examination of skeletal remains provided evidence of abuse that X-rays failed to reveal. They found that abused children had multiple areas of subperiosteal bone formation that were at different stages of healing. Interestingly, Kleinman et al. (1996) and Mandelstam et al. (2003) argued for high-resolution imaging that can identify pathological subperiosteal bone formation in living children who are suspected of being victims of abuse.

To diagnose physical abuse using the skeleton, several factors must be taken into account: the number of injuries, the cause or specificity of the injuries, and how old the injuries are (Kleinman et al. 1996). Patterns of fractures have been utilized to increase the diagnostic strength of determining abuse. For example, using a meta-analysis approach, Kemp et al. (2008) found that when looking at children with fractures, abused children are usually younger (80% are under eighteen months old), whereas in the non-abused, 85% of the children with fractures are over five years of age. Also, multiple fractures are more common in abused children than in non-abused children; Kemp et al. found that multiple fractures and fractures of the ribs, humerus, and femur were common in abused children under three years of age. However, by looking at a sample of 189 battered children and 429 fractures, King et al. (1988) concluded that about half of children who experience abuse had multiple fractures, whereas the other half had only one fracture. Rib fractures, especially those in the posterior-medial plane, are common in abused children but rare in non-abused children (Dwek 2011; Kemp et al. 2008; Mandelstam et al. 2003). Cranial injuries are perhaps the most difficult to diagnose, even though head injuries are the leading cause of death in abused children under two years of age (Rubin et al. 2003; Steyn 2011). Rubin et al. recommended that children at risk for abuse who have head injuries should be examined for other signs of abuse. Walker et al. (1997) found that the physical anthropological perspective can help determine the cause of cranial fractures. In fatal child abuse cases, Walker et al. found that linear cranial fractures were common; for example, in a case of an eleven-month-old, the baby had a Y-shaped fracture on the skull that was fifteen millimeters long and two regions of new bone formation on the parietals. Another case of abuse included a three-year-old with a linear fracture from the foramen magnum to the lambdoidal along with new bone formation near the fracture. Cranial injuries from falls tend to be more complex and spiral. Femoral fractures, especially in children too young to walk, occur more frequently in abused children than in non-abused children, but in older children, limb fractures are more likely to be accidental, which demonstrates the importance of taking age into account in determining the cause of injury (Kemp et al. 2008; Kleinman et al. 1995). Dwek (2011) suggested that fractures of the metaphyseal corners, posterior-medial ribs, sterna, scapulae, and vertebral spinous processes are the most diagnostic fractures of abuse. However, Dwek mentioned that rickets as a cause of these fractures needs to be eliminated. Other complicating factors also need to be assessed.

In the 2011 report of the National Child Abuse and Neglect Data System, it was estimated that about 1% of minors were abused in the United States and 17.6% of abused children experienced physical abuse. Although a quarter of fractures in babies less than a year old were a result of abuse, many other health issues can mimic abuse fractures (Egge and Berkowitz 2010). For example, osteogenesis imperfecta leads to bone fractures, but the rate of this disease is between one in ten thousand and one in twenty thousand; thus, when a fracture is found in children under three years of age, there is a twenty-four times greater chance of abuse than osteogenesis imperfecta (Egge and Berkowitz 2010). A forceful delivery of the child during birth and a variety of nutrient deficiency diseases (e.g., **scurvy** and rickets) can be mistaken for abuse. Copper deficiency can also predispose children to fractures; copper deficiency is common in individuals who were born premature, but the symptoms, which include pale skin and impaired growth, do not appear until six to twelve months of age (Egge and Berkowitz 2010). Femoral fractures in infants are likely the result of physical abuse; Grant et al. (2001) reported that between 12% and 29% of physically abused children have femoral fractures. However, other causes of femoral fractures should be ruled out. For example, Grant et al. wrote of two cases of femoral fractures as a result of Exersaucer baby walkers that enable babies to bounce around while being stationary.

Sports

The NIH (2013) reported that in 2009 there were thirty-eight million US children and teens involved in organized sports. In a single year, somewhere between 1.9 million and 2.6 million children between the ages of five and eighteen will visit an emergency room in the United States for a sports-related injury (Adirim and Cheng 2003). These numbers under-estimate the frequency of injuries in youth since many young athletes will go to their primary care physician for sports-related injuries rather than go to the emergency room, and many will not go to any doctor at all. Bach and Shilling (2008) estimated that of the more than thirty million children and teens involved in organized sports in the United States, 3.5 million will seek treatment for overuse or **chronic** fatigue.

The Canadian Hospitalization Injury Reporting and Prevention Program (CHIRPP) is a computerized information system that collects and analyzes data on injuries to people who visit emergency rooms; it has provided a wealth of information on trends of childhood injuries in Canada. Babul et al. (2007), for example, used CHIRPP to determine that a quarter of emergency visits for individuals under nineteen years of age were the result of sports-related injuries. Nearly eight out of ten of these emergency visits for sports-related injuries did not affect bones. Males were more likely than females to be injured, and the top sports that resulted in emergency room visits were cycling, basketball, and soccer. Interestingly, Babul et al. (2007) found that in most sports, non-organized (children just playing) sports led to more injuries, but in soccer and ice hockey, organized sports led to more emergency visits. Also using CHIRPP, Keays et al. (2006) found that when children experienced minor injuries they were more likely to experience injuries again, whereas major injuries reduced risk of future injuries likely because the children became more cautious.

The increase in sports-related injuries is a result of sports training for children and teens having become more intense in the last decade. Training centers that have been popping up since the 2000s in affluent neighborhoods have promised parents that their child will be the next sports celebrity (Stinchfield 2008). Children and teens are now often practicing year-round on a single sport, which increases risk of overuse injuries (Adirim and Cheng 2003; Bach and Shilling 2008). Some of the increase in sports in Western countries is a reaction to the increase in childhood obesity.

Sports injuries are most common between the ages of five and fourteen years old. Some sports are more likely to cause injury than others. Football has the highest injury rate, with between 41% to 61% of children and teens who engage in football experiencing injury (Adirim and Cheng 2003). Wrestling and gymnastics are the next most dangerous sports in relation to injury rates. Yet even low-risk sports, such as soccer and baseball, have injury rates between 7% and 18% (Adirim and Cheng 2003). The most common anatomical sites for injury include the ankle, knee, hand, wrist, elbow, and shin. These sites are evolved for specific motions, and sports may counter those motions; for example, the knee is evolved for flexion and extension, but some sports will cause more torsion and thus tear the ligaments at the knee and shin.

The most common types of sports-related injuries are musculoskeletal and involve either sprains or strains. Fractures, growth plate injuries, dislocations, and overuse or repetitive stress injuries are also common. Some injuries that were only previously found in adult athletes are now found in children; for example, Tommy John ligament injury, which is an overuse pitching injury that usually requires surgery, was previously found in athletes in their twenties and thirties, while now it has been documented in teen athletes. Overuse injuries in the United States are four times more common now than they were five years ago (Bach and Shilling 2008).

LITTLE LEAGUE ELBOW AND OVERUSE INJURIES. There are an estimated two million children in Little League baseball; a quarter of those playing will experience elbow pain (Klingele and Kocher 2002). The pain may be the result of a fracture, an avulsion, or a dislocation, but the most common confounding factor of all the injuries involves overuse of the elbow (Klingele and Kocher 2002). Overuse can lead to osteochondritis dissecans (OCD), which is a type of overuse injury in which cracks appear in the cartilage, but it can also lead to more severe fractures. In a season, Little League pitchers throw almost twice as many times as professional pitchers (Smith 2006). Furthermore, children tend to throw only with their arm, which increases the risk of injury, rather than utilizing their trunk, hips, and legs (Smith 2006). Due to the increase in throwing and the misuse of arm force, children and teens are at risk from what is known as Little League elbow. Little League elbow was first described in 1960 (Klingele and Kocher 2002). It is an injury in which the medial epicondyle fractures due to repetitive stress and micro-trauma. The elbow has six ossification centers, and the last of these locations to fuse is the medial epicondyle, which occurs at around fifteen to sixteen years of age. The medial epicondyle, in part because it is the last to fuse, is the weakest site in the elbow, and thus most at risk for injury (Klingele and Kocher 2002). In X-ray evaluations of Little League elbow, the indicators are subtle; there may be a slight difference in the arm's angle, a doctor may see some **osteophytes** or loose bone, or **chondromalacia** (poor formation of the cartilage) may also occur (Klingele and Kocher 2002). However, the most indicative signs of Little League elbow are pain and restrictive movement of the elbow. Although many Little Leaguer elbows get better with rest and without intervention, the long-term consequences may include growth issues and osteoarthritis of the elbow (Stinchfield 2008).

It is important to note that Little League elbow can occur during other activities. For example, anthropologists have reported that subadults in past populations have also experienced upper limb trauma related to activities, such as a seventeenth-century Canadian child with a "Little Leaguer's elbow" injury who may have received it from throwing hunting implements or from other activities (Glencross and Stuart-Macadam 2000).

Although Little League elbow has received the most attention, other overuse injuries are common in baseball too. Overuse injuries or repetitive motion injuries are usually visible in X-rays as hairline fractures; these may eventually lead to larger fractures. About half of all sports injuries in children and teens are overuse injuries (Kuzma 2012). Overuse injuries outside of baseball include shin splints in runners, tendinitis in tennis players, and wrist injuries in gymnasts (Kuzma 2012).

SHOULDER DISLOCATION. Football players face many different injuries; they experience strains, pulled muscles, tears, fractures, and dislocations (NIH 2013). One of the common dislocations experienced by football players in their youth is a **glenohumeral** dislocation (Kocher et al. 2000). The glenohumeral joint is where the upper humerus meets the scapula's glenoid. The shoulder joint is the most often dislocated joint in humans (Kocher et al. 2000). Throughout human evolution there has been a selection for a more flexible, and thus less stable, shoulder joint. This flexibility allows humans to use their arms in ways that other animals who use their shoulder in locomotion cannot. The hemispherical humeral head rests against a slightly concave **articular surface** on the lateral side of the shoulder joint. The shoulder is held together by fibrous tissue and is not strongly supported by the osteological construction, which differs from the hip joint where the femoral head slides tightly into the pelvis's acetabulum. Therefore, humans rarely dislocate their hip but dislocate their shoulder often. Figure 4.3 displays the difference be-

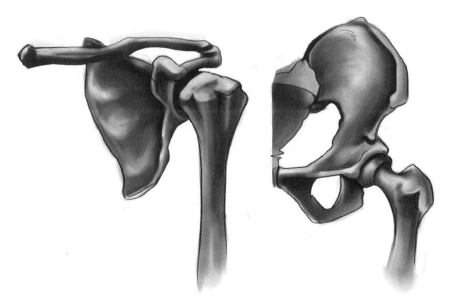

Figure 4.3. Shoulder joint compared to the hip joint. The skeletal composition of the shoulder joint (left) is looser, which allows for more flexibility than the hip joint composition (right), which has been selected for stability. Illustration by Vanessa Corrales.

tween the loose skeletal configuration of the shoulder joint compared to the tight skeletal configuration of the hip joint. Football players experience shoulder dislocation even more frequently than non-players; their strength and the throwing of the ball can result in dislocation, but more frequently tackling likely does the harm. When the shoulder is not fused completely, then the dislocation may be a growth plate injury instead. Regardless, injuries of the glenohumeral joint often recur even with surgery and can lead to osteoarthritis later in life (Kocher et al. 2000).

GROWTH PLATE INJURIES AND AVULSIONS. One in six sports-related fractures in youths are growth plate injuries (Schwager 2010); these can involve avulsions or occur without avulsions. Knees are frequently affected by these injuries, and the result may be permanent growth arrest or crooked growth if nerves and blood vessels have been damaged (Schwager 2010). Growth plate injuries, which are sometimes called Salter-Harris injuries, range from a simple separation of the epiphysis from the metaphysis while the growth plate remains intact (which is considered type I) to the fracture breaking off of the epiphysis, the physis, and the metaphysis (which is considered type IV). The rarest form is type III, where a part of the epiphysis is separated from the growth plate and the metaphysis. The most severe is type V where the injury involves the crushing of the end of the bone and the growth plate. Most commonly, the epiphysis and growth plate are separated, but the metaphysis and growth plate are not (type II).

Males are at greater risk of growth plate injuries than females since they tend to mature more slowly than females (Schwager 2010). And about 10% of children and teens who experience a growth plate injury will require surgery or manipulation (Schwager 2010). The most common cause of growth plate injuries are sudden forces, such as a nasty fall, but a third of such injuries are sports related (Schwager 2010).

Avulsions come in two forms: sprains and strains. Sprains are injuries to a ligament, whereas strains are injuries to a muscle or tendon. They both involve injuries to fibrous cords and relate to muscle pulls (NIH 2013). As mentioned earlier, muscle and tendon pulls are frequent in children and teens because they exert extra strength and energy through muscle use, but the attachment sites are not strong enough for the forces (Balassy and Hörmann 2008; Donnelley et al. 1999). Many of these injuries may not show up on X-rays until the tendon or ligament has ossified (Schwager 2010). Although most avulsions can heal on their own with rest, some avulsions disrupt growth. Avulsions that involve tensile forces tend not to result in growth disruption (Caine et al. 2006). This may be because the site experiencing the injury is evolved for tensile forces; muscles and their connective tissue tend to pull at bones to enable movement. Compression, however, does affect growth; these compressions often are in connection to growth plate injuries.

Soccer players are at high risk for tendon tears and kneecap damage as a result of these sprains and strains (Stinchfield 2008), but football players, basketball players, track and field runners, and gymnasts also face sprains and strains. One common type of muscle pull is Osgood-Schlatter disease. Osgood-Schlatter disease can result in a painful bump below the patella, accompanied by pain during running or walking, and it usually improves with rest. Osgood-Schlatter disease, which affects up to 20% of adolescent athletes, is the result of the patellar tendon that attaches to the quadriceps being pulled away from the bone too often and with too much force for the attachment site. Using a sample of nearly a thousand children aged from twelve to fifteen years old in the Brazilian school system, de Lucena et al. (2011) found that athletic students, regardless of sex, were more likely to experience Osgood-Schlatter disease than non-athletes. Some early evidence of Osgood-Schlatter disease comes from DiGangi et al. (2010), who found a skeleton with Osgood-Schlatter disease from East Tennessee dated between AD 900 and 1600; they ruled out other diseases since just one bone was affected. The pulled muscle may have been a result of activities, including running or walking long distances.

Obesity

The rate of childhood obesity has tripled since the previous generation (Edmonds and Templeton 2013). Rates of obesity in people under nineteen years of age depend on locations, but Pollack (2008) has stated that one in six children between the ages of two and nineteen years of age are obese in the United States. Even in developing countries, childhood overweight and obesity issues arise; for example, Egypt, Morocco, Uzbekistan, Algeria, and Zambia all have as many overweight preschool children as underweight preschool children (de Onis and Blössner 2000). Overweight and obese children have earlier epiphyseal fusion and higher bone mineral content; thus, one may logically conclude that obese children are less likely to experience fractures or epiphyseal injuries. Yet data from emergency rooms present a different picture. Overweight and obese children face an increase in slipped capital femoral epiphyses, Blount's disease, and fractures.

Slipped capital femoral epiphyses occur when there is a separation of the ball of the hip joint at the superior growing end of the epiphysis (Sankar et al. 2011). In other words, it is a type of femoral head dislocation when the epiphysis has not fused. Children with slipped capital femoral epiphyses have widened growth plates as a result, and an increase in the hypertrophy zone, which is the zone that enlarges the cartilaginous cells prior to calcification. Thus, the epiphyseal region is less organized and weaker, which can cause slippage and shear forces on the hip joint (Walker et al. 1997). Additionally, shearing forces are exaggerated on the hip joint as a result of excess weight

(Pollack 2008). Slipped capital femoral epiphysis typically occurs between ages eleven to fifteen and is more frequent in boys than girls. Must and Strauss (1999) found that femoral head dislocations occurred in 3.4 out of one hundred thousand children, but that 50% to 70% of these children are obese. Two-thirds of the **bilateral** (occurring on left and right sides) slipped capital femoral epiphyses occur in obese children (Must and Strauss 1999). These injuries can be detected in X-rays, but they are also preliminarily recognized through symptoms. Symptoms of slipped capital femoral epiphyses include difficulty walking, knee and hip pain, hip stiffness and restricted movement, and outward turning of the leg (Sankar et al. 2011). In order to treat slipped capital femoral epiphyses, surgery is required to stabilize the joint and prevent excessive wear on the femoral head (Sankar et al. 2011). Regardless of the treatment, the risk of hip arthritis later in life is increased.

Another frequent injury in obese children is Blount's disease; Blount's disease is the bowing of tibiae as a result of excess weight on the growth plates (Canale 2007; Pollack 2008). Figure 4.4 illustrates Blount's disease bowing. One possible reason for Blount's disease is that cartilage can start to enter the metaphysis (Pollack 2008). Bowing of tibiae in very young children will straighten out with growth; bowing can also be the result of rickets. However, in obese children the

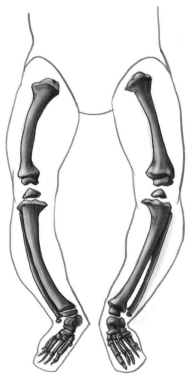

Figure 4.4. Blount's disease. An illustration of the bony changes that occur in Blount's disease, which may result from excessive weight. Illustration by Vanessa Corrales.

bowing seems to occur right below the knees, and it rapidly gets worse (Canale 2007). X-rays can help to diagnose Blount's disease. Usually braces are used to straighten the legs, and weight loss is recommended (Canale 2007). When not treated, Blount's disease can cause in-toeing, which is sometimes called pigeon-toeing, and arthritis later in life. Even though Blount's disease is more common in heavier children, Pollack (2008) commented that it is unknown whether obesity causes Blount's disease or whether the inactivity of children with Blount's disease causes obesity; it is likely a feedback loop.

Fractures of the upper limb are more common in obese children than in non-obese children. Fornari et al. (2013) reported that the risk of upper limb fractures is 1.7 times higher in obese children than in children with normal BMIs. The elbow region is especially at risk in obese children. Nearly two-thirds of arm bone fractures occur at the elbow in obese children. These fractures are low-energy fractures, which suggest they usually occur during non-sport activities (Fornari et al. 2013). In a sample of nearly a thousand children, Fornari et al. (2013) found that lateral epicondylar fractures increased with BMI. The lateral epicondylar fractures may be the result of fully extending the arm when falling; figure 4.5 illustrates a typical lateral epicondylar

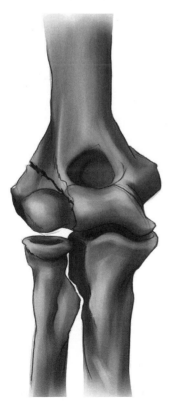

Figure 4.5. Lateral epicondylar fracture. Sometimes heavy children can have lateral epicondylar fractures as a result of tripping and catching their fall. Illustration by Vanessa Corrales.

fracture. Additionally, obese children fall more frequently due to gait problems and fall with more force. Although in most cases fractures can be treated and heal, fractures increase the rate of arthritis in later life. They also expose children to complications from treatment and may reduce activity levels during their healing period, which can exacerbate weight problems.

CONCLUSIONS

It is difficult to ascertain what changes have occurred in regard to childhood injuries from thousands of years ago to the present. The main difficulty arises from a lack of information about children and their injuries in prehistoric populations; their bones did not preserve well, and childhood injuries that have healed may not be visible on adult remains. Sometimes antemortem and postmortem damage on bones in young individuals is difficult to distinguish. Thus, for the most part, bioarchaeologists focus on adult bones and injuries of adults. A few case studies have revealed some possible links between injuries and activity patterns, but the information does not reveal patterns about the past. What is evident in the present is that childhood injuries are often the result of sports-related injuries or obesity; these two factors seem to show once again how the modern world is a world that is defined by extremes.

CHAPTER 5

Back Pains

Roughly 80% of people experience back pains in their life (Jensen et al. 1994; Peng et al. 2009). At any time, 15% to 20% of the adult population in the United States is experiencing lower back pain (Peng et al. 2009). Lower back pain is the second most diagnosed medical condition, while the common cold is the most common reason people seek medical attention (Jensen et al. 1994; Peng et al. 2009). Lower back pain is the most common reason for disability and work-time losses (Kalpakcioglu et al. 2009).

Although back pains, and especially lower back pains, are extremely common, the causes of these pains are extremely difficult to determine; pinpointing the exact source of the pain is nearly impossible (Ehrlich 2003). Specific causes, such as falls, trauma, or osteoporotic fractures, account for less than 20% of back pains (Ehrlich 2003). Ways to determine causes include examining vertebral anatomy, assessing whether specific physical traits (such as **hernias**, slipped discs, and fractures) correlate with pain, and looking at the demographics of back pain.

VERTEBRAL ANATOMY OF A BIPED

Our vertebral column is modified from a quadruped's vertebral column. All other higher-order primates, which include New World monkeys, Old World monkeys, Asian apes, and African apes, move about on four limbs; that is, they are quadrupeds. Thus, the last common ancestors with humans' closest cousins, the apes, were likely quadrupeds, and the human vertebral column was initially adapted to **quadrupedalism** rather than **bipedalism** (using two limbs to move around).

The vertebral column had to change to be effective for bipedality. Chimpanzees are good comparative animals to help us understand the anatomical modifications that were necessary to become a bipedal primate. A chimpanzee's vertebral column is not evolved as a weight-bearing column as ours is; rather the job of a quadruped's vertebral column is more like a suspension bridge in that it is made to move on the horizontal plane. The action of arching and stretching the back on the horizontal plane allows for greater flexibility in the middle and upper back and allows for an increase in stride length when compared to human back movements. Also, each vertebra in a quadruped's spine is about equal in size, and the intervertebral discs (which are fibrocartilaginous structures) are thin and cover the entire superior and inferior surface of the **vertebral body**. The thin intervertebral discs coupled with the lack of size differences in the vertebrae of apes results in a fairly straight spine, which has much flexibility in the middle and upper regions.

In a quadruped, the center of gravity is below the center of the torso; when they move forward, their mass is constantly still over their center of gravity. In humans, however, the center of gravity is at the center of the pelvis, and thus when a leg swings out to walk, it is out of the center of gravity, which makes falling over more likely. In order to prevent falling, the human backbone has curved, which enables the body's mass to still be over the center of gravity even when walking. The human spine has an S-shaped curve that pushes some of the body's weight forward into the pelvis region. The lumbar curve, which is in the lower region, allows for greater flexibility in the lower back to rotate the hip and thereby also keep the body's weight above the center of gravity. The curve is attained by the evolutionary modification of larger lumbar vertebrae; there has been evolutionary selection for an increase in size in the lower vertebrae (lumbar vertebrae) compared to the upper vertebrae (the cervical and thoracic vertebrae). In apes, all vertebrae are about equal in size. Lumbar flexibility is also obtained by increasing the number of lumbar vertebrae from four to five. Apes have four lumbar vertebrae and six sacral segments; the sacrum is fused together in most animals (including all primates) and thus is less flexible than the rest of the spine. Humans have five lumbar vertebrae and five sacral segments; but these numbers are not fixed, and for humans, having a sixth lumbar vertebra, for instance, is not uncommon.

Another component of the changes in the human vertebral column relates to weight bearing. Being a biped requires that the weight of the upper body be carried by the lower body. Thus, changing the spine into a weight-bearing structure, rather than a stride-increasing suspension bridge structure, includes increasing vertebral size from the small cervical to the larger thoracic and the large lumbar vertebrae. A larger size will experience less weight on any one point; thus, the weight is more evenly distributed.

With each step, the biped also experiences more shock due to the weight and the increased flexibility of the lumbar region, and thus shock absorption becomes an essential evolutionary modification to make bipedality feasible. Increases in the size and structure of the intervertebral discs allow for the reduction in shock felt when walking and running. Humans have thick and cushioning intervertebral discs. The main purpose of a quadruped's intervertebral discs is to allow for more flexibility; thus, they act more like the cartilage of moveable joints. The shock-absorbing thick intervertebral discs, however, do lose height and flexibility over time, and they are more likely to become herniated than the thin chimpanzee discs.

BACK PAIN DEMOGRAPHICS

The vertebral changes are not perfect; evolution, after all, modifies existing structures and does not start from scratch. The frequent hernias, slipped discs, and fractures are likely the result of the imperfection of the spinal column's adaptations from a suspension bridge structure to a weight-bearing and shock-absorbing structure.

However, not all osteologists blame evolution for back pain. For example, Ehrlich (2003) argued that evolutionary baggage is not the cause of back problems; he cited links to workplace dissatisfaction and noted the lack of correlation between structural issues, such as hernias, and pain to support his claim that back pain is more highly influenced by psychology than morphology. One way to determine whether any ailment has an evolutionary source is to examine who the afflicted individuals are; evolutionary causes should affect people across cultures. Furthermore, sex and age differences should be uniform from one population to another; in other words, who gets back pains in any particular culture should be identical to who gets back pains in any other particular culture. Stating that there may be evolutionary reasons for back pain, however, does not mean that some individuals may not be more prone to back pain than others; for instance, anatomical variation in an individual that may include having an additional lumbar could increase the individual's chance of developing back pain. With those caveats, looking at the demographics of back pain should still help us decide whether the pain is a result of evolutionary baggage or not.

In general, back pains are experienced more frequently by adults than children; the peak back pain frequencies occur between ages forty-nine and fifty-nine years (Ehrlich 2003). In most of human evolution, individuals were unlikely to live this long. Older adults are likely a recent phenomenon in human evolution (Caspari and Lee 2004); thus, the evolutionary modifications would not have led to back pain until relatively recently. Young athletes are the exception to this pattern; for instance, young gymnasts, weight lifters, and soccer players often experience "adult" back pains (e.g., Commandre et al. 1988; Iwamoto et al. 2005; Reitman et al. 2002). One may argue that athletes mimic early humans' behavior better than non-athletes, but many sports involve actions that are not seen in hunter-gatherers.

There appear to be no ethnic or sex differences in back pain, but back pain does increase with obesity in both children and adults (Ehrlich 2003). De Sá Pinto et al. (2006) found that in a study of children between the ages of eight and fourteen years, obese children had an increased risk of lower back pain compared to non-obese children. Other studies linked lower back pain in children to obesity and sedentary behaviors, such as watching TV and playing on computers (Kristjansdottir and Rhee 2002; Malleson and Clinch 2003). Obesity is a recent phenomenon, and thus early humans were less likely to experience back pain.

The demographics of back pain suggest that it is prevalent as a result of evolutionary baggage combined with Westernization, but there may be other factors at play, such as vertebral pathologies that osteologists find in skeletal samples. There are three common types of vertebral traits that can be found in the bioarchaeological record that are also diagnosed in clinical settings: **Schmorl's nodes** (hernias), **spondylolysis** (fractures), and **spondylolisthesis** (slipped disks). Degenerative disc disease is usually treated as osteoarthritic by anthropologists and will be addressed in chapter 6. Other vertebral problems, such as Modic changes (which are also known as vertebral end plate changes), are not recorded in skeletal remains. Back curvature disorders, such as kyphosis, **scoliosis**, and lordosis, are usually considered rare. They may be discussed in connection to other back anomalies but are not usually considered the cause of most individuals' back problems.

VERTEBRAL PATHOLOGY DETECTION

In clinical settings, hernias, slipped discs, and fractures are diagnosed using CT scans, MRIs, or PET scans. PET is an abbreviation for positron emission tomography, and it is a nuclear medical imaging technique that produces a three-dimensional image of both bone and soft tissue. Some clinical studies have questioned the use of simple X-rays; Ridgewell (2003) suggested that CT scans, MRIs, and PET scans are good tools for diagnoses, but that regular X-rays will miss most vertebral anomalies, such as slipped discs, spondylolysis, and hernias. Simple X-rays, actually, are not considered useful in most cases, but these X-rays are still common in chiropractors' offices (Ernst 2008). Although imaging techniques are often used to determine spinal disorders, sometimes assessments are made without imaging techniques. For example, palpation may also be utilized (Ehrlich 2003; Ernst 2008). Kalpakcioglu et al. (2009) have reported that determining spondylolisthesis with clinical evaluation (and not imaging) is effective and efficient; however, Collaer et al. (2006) reported high error rates in clinical evaluation of back pains.

In skeletal records, the physical traits of the spine are investigated macroscopically. Most frequently, anthropologists will just look at the backbones to determine whether a hernia, slipped disc, or fracture has occurred. Sometimes X-rays will be utilized to determine whether a fracture occurred that has healed and can no longer be seen from the bone examination. When looking for slipped discs, anthropologists often require multiple vertebral bones to be placed in anatomical position to determine which vertebral body would have been pushed or slipped forward; assessing slippage without the entire verte-bral region (i.e., lumbar, thoracic, or cervical) is difficult and may cause an-thropologists to miss the subtle changes in the bone that occur with slippage.

SCHMORL'S NODES

In 1927, pathologist Christian Georg Schmorl described hernias that leave a mark on vertebral bodies (Kyere et al. 2012). Schmorl's nodes are herniations of the nucleus pulposus (which are jelly-like structures) of the intervertebral discs through the cartilaginous end plate and into the cancellous bone of the vertebral body (Burke 2012; Kyere et al. 2012). Schmorl's nodes occur most frequently in the lower thoracic and lumbar vertebrae. They appear as dents or impressions on the upper or lower vertebral body. Figure 5.1 displays a Schmorl's node. Figure 5.2 displays a less common cervical hernia. Schmorl's nodes population rates vary from 8% to 80% (Burke 2012; Kyere et al. 2012).

Schmorl's Nodes Risk Factors

Many clinicians and osteologists assume that Schmorl's nodes are at least in part a result of trauma from activities. For example, Dar et al. (2010) exam-ined skeletons from the Hamann-Todd Collection and found that Schmorl's nodes showed no sex, race, or age differences, and thus they assumed that

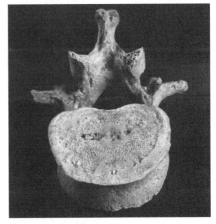

Figure 5.1. Schmorl's node. In the thoracic vertebrae of this eighteen- to thirty-year-old female, Schmorl's nodes are present. Here we see one on the posterior (left) and superior (right) body of the vertebrae. Photographs by Daniel Salcedo.

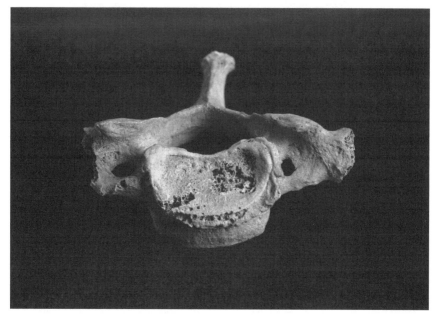

Figure 5.2. Cervical hernia. Hernias in the upper vertebrae are less common than the typical thoracic or lumbar hernias; this hernia of the cervical vertebra was found in a twenty-five- to thirty-five-year-old male. Photograph by Daniel Salcedo.

Schmorl's nodes are more likely a result of trauma. Burke (2012), using a military sample of 172 individuals, found nearly three-quarters of her sample had Schmorl's nodes, which were attributed to trauma. Anthropologists also associate Schmorl's nodes with repetitive strains. For example, Wentz and De Grummond (2009) looked at a Ukrainian sample dated to 225 BC and attributed Schmorl's nodes to horseback riding among the warrior class. And Weiss (2005) found that Quebec prisoners of war who carried heavy loads were more likely to have Schmorl's nodes than British Columbian Native Canadians. Kelley (1982) established that prehistoric hunter-gatherers of the Indian Knoll in Kentucky developed Schmorl's nodes earlier in life than individuals from the Hamann-Todd twentieth-century collection; Kelley attributed this difference in Schmorl's nodes onset to a less strenuous life in the Hamann-Todd Collection than in the hunter-gatherers.

Although no sex differences in Schmorl's nodes were found in the Hamann-Todd Collection, research on bioarchaeological samples has produced sex differences. Often authors who find sex differences in Schmorl's node frequencies attribute the difference to sexual division of labor. For example, Jiménez-Brobeil et al. (2010) examined a Bronze Age (1800 to 1300 BC) sample from Spain and concluded that the high frequency of Schmorl's nodes in males from this population was likely a result of a strenuous lifestyle that included min-ing, livestock care, and sword fights. The females had fewer nodes and were thought to have taken care of children and engaged in less strenuous domestic activities. Novak and Šlaus (2011) examined medieval skeletal remains from Croatia and found that males had five times the Schmorl's node frequency compared to females; they attributed these differences to activity patterns, with males plowing, hoeing, and doing carpentry while females wove, cleaned, and cared for children. Other studies, such as Üstündağ (2009) and Wentz and De Grummond (2009), also found more Schmorl's nodes in males than in females. Even in cadaver studies, higher frequencies of Schmorl's nodes have been found in males than in females (Kyere et al. 2012). In order to conclude that the Schmorl's nodes are caused by activity, a reverse sex difference needs to be discovered. If regardless of the population males have more Schmorl's nodes than females, then the sex difference is likely to be biological.

Plomp et al. (2012) cautioned against assuming that Schmorl's nodes are a result of trauma or repetitive stress. Using a skeletal sample from England, Plomp et al. examined the frequency of Schmorl's nodes in relation to vertebral morphological variation; they found that pedicle shape, posterior margin of the vertebral body, and vertebral body size influences the presence or absence of Schmorl's nodes. One indication that trauma is not necessarily the cause of Schmorl's nodes is the high heredity factor of Schmorl's nodes. Using a twin study design with 516 twins (150 monozygotic and 366 dizygotic), Williams et al. (2007) found that Schmorl's node heritability is between 70% and 80%; they also found a relationship between Schmorl's nodes and lumbar disc diseases.

Schmorl's Nodes and Pain
Another assumption that should be avoided is that Schmorl's nodes are related to back pain. Kyere et al. (2012) reported in a meta-analysis that Schmorl's nodes are associated with pain when they first occur, but the pain usually is resolved spontaneously. Others, such as Peng et al. (2003) and Hasegawa et al. (2004), have found that surgery reduced the pain caused by Schmorl's nodes. Yet other scientists are not convinced of the link between trauma, pain, and Schmorl's nodes. For example, Jensen et al. (1994) found that nearly all cases of vertebral disk bulging and compressions were **asymptomatic**; only hernias that displayed extrusion into the vertebrae were more likely to be found in pained individuals than in non-pained individuals.

SPONDYLOLYSIS AND SPONDYLOLISTHESIS
Both spondylolysis and spondylolisthesis are not difficult to ascertain in the skeletal record. Spondylolysis is a vertebral fracture; *lysis* means break in Latin. Spondylolisthesis is a slipped vertebrae; *olisthesis* means slip in Latin.

Although both of these pathologies usually occur in the lower back, they can also occur in the cervical region (e.g., Dean et al. 2009).

Spondylolysis

Spondylolysis is a fracture that causes the separation of the vertebral arch from the vertebral body (Merbs 1996). Specific demographic and morphological patterns of spondylolysis have been found that transcend culture and time, which suggests the cause of spondylolysis may lie in the evolutionary shift from quadrupedalism to bipedalism. For example, the fifth lumbar is the most frequently affected bone, followed by the fourth lumbar (Merbs 1996; Ward et al. 2010). The most common pattern found in cases of spondylolysis, which is referred to as "classic spondylolysis," is defined as occurring in the fifth lumbar with complete bilateral separation (Merbs 1996). Classic spondylolysis is shown in figure 5.3. Also, spondylolysis does not usually occur in individuals who are too young to walk, which has led to the suggestion that these fractures

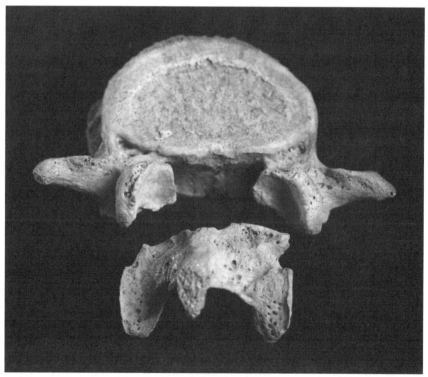

Figure 5.3. Classic spondylolysis. Displayed is a classic, bilateral case of spondylolysis from a thirty-nine- to forty-four-year-old male. In a living individual, these two parts of the vertebra are held together by soft tissue, but in the skeletal collection they are found separated. Photograph by Daniel Salcedo.

may be an adaptation to bipedality rather than an injury that causes pain (Merbs 1996; Sonne-Holm et al. 2007). To further support the link between spondylolysis and bipedality, Eisenstein (1978) noted that spondylolysis is *Homo sapiens* specific; that is, it does not occur in the other living primates.

Sex differences also seem universal; males are more likely to be afflicted by spondylolysis than are females (Merbs 1996; Weiss 2009). Kalichman et al. (2009) found in a study of three generations of families in the Framingham Heart Study that the sex ratio of spondylolysis was 3.3 males to 1 female, and Sonne-Holm et al. (2007) found in a study of over four thousand Danish individuals that males had twice the rate of spondylolysis as compared to females, except in the fourth lumbar where females had more spondylolysis than males. Interestingly, Sonne-Holm et al. also found that an increase in BMI correlated with fourth lumbar fractures, but not with the more common fifth lumbar fractures. An exception to the sex difference was found in Nebraskan prehistoric skeletal remains; females had more spondylolysis than males (Reinhard et al. 1994). This reverse sex difference was attributed to activities, such as home building, in the Nebraskan Natives.

However, although these nearly universal patterns exist, rates of spondylolysis vary dramatically from one population to another; in some populations rates are as low as 5%, such as the Hamann-Todd Collection (Willis 1923), whereas in other populations, such as the Inuit, half of the population may be affected (Merbs 1995). Additionally, in clinical studies young adults are most likely to be afflicted, but in archaeological samples middle-aged adults have higher rates of spondylolysis than any other age group (Merbs 2002; Shrier 2001). Due to these differences and the rare cases of multiple spondylolysis and non-lumbar spondylolysis, anthropologists and clinicians have been examining causes beyond bipedality for spondylolysis. In addition, questions regarding whether spondylolysis causes pain are being asked by clinicians and anthropologists.

SPONDYLOLYSIS RISK FACTORS. Athletes and especially young athletes are at greater risk of spondylolysis than non-athletes. Soccer players (Álvarez-Díaz et al. 2011), rugby players (Iwamoto et al. 2005), gymnasts (Commandre et al. 1988), rowers (Rumball et al. 2005), weight lifters (Risser 1991), and players of sports with bats, like baseball and cricket (Ruiz-Cotorro et al. 2006), all seem to have elevated rates of spondylolysis. Although most of these cases follow the pattern mentioned previously (e.g., the fifth lumbar is most frequently affected, and bilateral fracture is more common than asymmetrical fractures), Sasa et al. (2009) reported on a rare case of cervical spondylolysis that occurred in a twelve-year-old judo athlete. Dunn et al. (2008) examined spondylolysis stress fractures as a result of sports and found that there are actually three stages of stress prior to the fracture; this implies that spondylolysis is not caused by trauma, but rather by repetitive stress. Anthropologists and clinicians have suggested that spondylolysis fractures are caused by hyperextension of the hip and lower back, and swinging a bat may cause torsion of the lower back. Those who treat young athletes have reported that spondylolysis does cause pain. Dunn et al. (2008) stated in their study of 156 teen athletes that nearly half of the lower back pains could be attributed to spondylolysis; however, nearly all of the bilateral fractures healed spontaneously.

Anthropologists have utilized the data from these clinical studies to infer that spondylolysis in past populations is a result of activity patterns. For example, Arriaza (1997) linked high levels of spondylolysis with the moving of stone pillars in Guam. Merbs (1983) found that Arctic populations' high rates of spondylolysis may be related to bending at the waist for activities and the posture used in kayaking. Still, many anthropologists also find that spondylolysis etiology is **multifactorial**, and even though activity may explain some of the population patterns, anatomical variation may explain some of it as well. Plus, some researchers using large databases of living peoples have not found a link between activity and spondylolysis; Sonne-Holm et al. (2007) using a sample of over four thousand individuals found that activities related to occupation did not correspond to spondylolysis.

For both clinical and anthropological researchers, examining anatomical variation is a popular way to study spondylolysis etiology. Aebi (2012) reported on a case study of a fifty-year-old male who had spondylolysis and underwent surgery for back pain that may have been related to the fracture; he found that transitional anomalies in the male may have led to the fracture. **Lumbarization** (which is caudal shift in vertebrae) and **sacralization** (which is a cranial shift in vertebrae) may increase risk of spondylolysis. Figure 5.4 shows lumbarization. Sacralization may lead to asymmetrical facets of the sixth lumbar and the sacrum; this asymmetry is visible in figure 5.5. Sacral segments are similar to lumbar vertebrae in lumbarization, and lumbar vertebrae are similar to sacral segments in sacralization. Merbs (1983) and Weiss (2009) found that with males, lumbarization correlated with spondylolysis in prehistoric populations in the Arctic and California. Other anatomical variations may also translate into a predisposition to spondylolysis; Bajwa et al. (2012) examined skeletal remains from the Hamann-Todd Collection and found that in a sample of over one thousand individuals, pedicle length was longer in individuals with spondylolysis than in individuals without spondylolysis. Also using the Hamann-Todd Collection, Masharawi et al. (2007) found that individuals with spondylolysis had a more trapezoid-shaped neural arch than non-afflicted individuals. Sonne-Holm et al. (2007) also reported that lumbar lordosis (i.e., pelvic inclination) was related to spondylolysis in both males and females. Anatomical variation is usually genetically determined, and Sairyo et

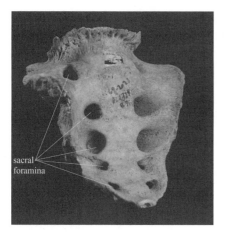

Figure 5.4. Lumbarization. In lumbarization, the sacrum consists of six segments, whereas normal sacra consist of five segments, which are marked by the sacral foramina. Ergo, a normal sacrum has eight sacral foramina, whereas a sacrum with lumbarization has ten sacral foramina. A case of lumbarization, as seen from the anterior (left) and posterior (right) from the bones of an adult male; lumbarization has been linked to spondylolysis in males. Photographs by Daniel Salcedo.

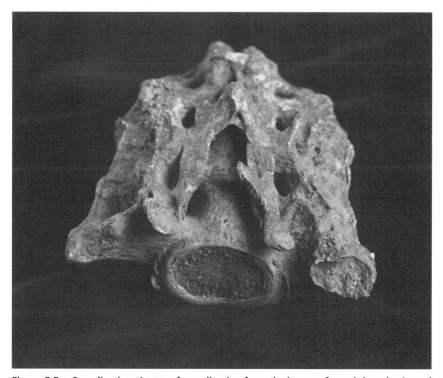

Figure 5.5. Sacralization. A case of sacralization from the bones of an adult male viewed from the superior sacrum with the posterior upward. Photograph by Daniel Salcedo.

al. (2009) reported that some individuals may have a genetic predisposition for spondylolysis stress fractures. Sairyo et al. found that spondylolysis occurs in families, and the rare second lumbar spondylolysis is found in both individuals of identical twins.

SPONDYLOLYSIS AND PAIN. The research on athletes suggests that spondylolysis causes pain (e.g., Dunn et al. 2008; Reitman et al. 2002; van der Heijden et al. 2007). However, since many of the athletic cases were discovered because the individual was in pain, it may be that many individuals have spondylolysis, but since they do not feel pain, they are never diagnosed. Some clinicians, such as Aebi (2012), have suggested that even if spondylolysis afflicts younger adults, the back pain may take a long time to develop. Others have reported the opposite; Brooks et al. (2010) used the US Health Insurance Portability and Accountability Act (1996) to examine data from CT scans and found that in a sample of 2,555, spondylolysis was prevalent in about 8% of the sample, but the presence of spondylolysis did not necessarily correspond to symptoms. It appears, according to Brooks et al., that developing spondylolysis may be painful, but once it has occurred, the pain subsides. Wicker (2008) reported that spondylolysis in grade school children does not correspond to back pain as it does in an adult; he added that the cause of spondylolysis is unknown and these fractures may be asymptomatic. And Sonne-Holm et al. (2007) did not find a relationship between pain and spondylolysis in a sample of 4,151.

Spondylolisthesis

Even if spondylolysis does not actually cause pain, it may lead to vertebral slippage, which is called spondylolisthesis. Determining the relationship between spondylolysis and spondylolisthesis has been difficult. Nevertheless, some large studies, such as Sonne-Holm et al.'s (2007) study of over four thousand individuals, found that although vertebral slippage is often age related, spondylolysis can often lead to spondylolisthesis.

Spondylolisthesis is recorded by clinicians on the Meyerding scale, which is based on grades I to V. These grades refer to the percentage of the vertebra that has slipped in front of or behind its adjoining vertebrae; grade I, for example, is when 25% or less of the slipped vertebral body is not in contact with its abutting vertebrae. The most severe grade (V) occurs when the slipped vertebra is completely off of the vertebrae above and below it. A vertebra can slip forward, which is called **anterolisthesis**, or backward, which is retrolisthesis. A narrowing of the central spinal foramen, which is called spinal **stenosis**, and compression of nerves may result from spondylolisthesis. These grades are relatively easy to determine using the imaging techniques discussed above, but in the bioarchaeological record, spondylolisthesis is a bit more difficult to assess.

In the bioarchaeological record, spondylolisthesis is usually recorded with spondylolysis. Mays (2006b), for example, examined spondylolysis and spondylolisthesis in tenth- to nineteenth-century AD British skeletal remains from Wharram Percy and found that spondylolysis rates were high in the population, but spondylolisthesis rates were low. She allowed for the possibility that the rates of slippage were low due to the difficulties in assessing slippage in dry bones; spondylolysis is easily diagnosed within each vertebra, but spondylolisthesis requires articulating backbones. Although it may be more difficult to assess spondylolisthesis in dry bones than it is to diagnose in living people, spondylolisthesis has been found in a variety of prehistoric remains. For example, Merbs and Euler (1985) reported on an Anasazi prehistoric Arizona case of a middle-aged female with multiple vertebral defects including both spondylolysis and spondylolisthesis. Edelson and Nathan (1986) found evidence of spondylolisthesis with spinal stenosis that may have compressed or hooked the nerve in remains from the Middle East dating to 1000 BC. O'Connor et al. (2011) found evidence of cervical spondylolisthesis that may have been related to **traumatic** force at the neck in a 2,500-year-old British male. Finally, the oldest case comes from a *Homo heidelbergensis* specimen dating to around half a million years ago; the individual had lumbar kyphosis and spondylolisthesis (Bonmatí et al. 2010).

SPONDYLOLISTHESIS RISK FACTORS. Spondylolisthesis is sorted into five different types: dysplasic (i.e., congenital), isthmic (i.e., related to stress fractures), degenerative (i.e., age-related slippage), pathological (i.e., slippage as a result of a second disease, such as cancer or Paget's disease), and traumatic (i.e., slippage as a result of an accident). The most common forms of spondylolisthesis are isthmic and degenerative. Isthmic spondylolisthesis risk factors include repetitive trauma and a family history. Sports players seem to be particularly prone to developing both isthmic and traumatic spondylolisthesis (Watkins and Watkins 2010; Wicker 2008). Van der Heijden et al. (2007), for instance, reported on a twenty-seven-year-old soccer player with gluteal pain that was coupled with spondylolisthesis and spondylolysis at the fifth lumbar.

Some congenital causes of spondylolisthesis include kyphosis and **spina bifida**. Kyphosis, which is curvature of the spine, can increase the risk of any type of spondylolisthesis (Vialle et al. 2007). In females, lumbar scoliosis has been shown to increase the risk of vertebral slippage (Wicker 2008). Sonne-Holm et al. (2007) noted pelvic inclination expressed as lumbar lordosis can lead to spondylolysis and in turn predispose individuals to spondylolisthesis. Additionally, spina bifida, a congenital deficiency of spinal column fusion, and other congenital pars interarticularis defects have been found to correlate with spondylolysis and spondylolisthesis in the clinical record (Mays 2006b).

Although these congenital spondylolisthesis causes are rare, less severe anatomical variances, which are likely genetic, can lead to spondylolisthesis.

Spondylolisthesis, which has been known since the early nineteenth century, was assumed initially to be a result of gravity and pelvic tilt (Merbs 2001). This theory was then dismissed because it was thought that an intact neural arch would prevent slippage; now, once again, osteologists accept the possibility that bipedality coupled with anatomical variance can predispose individuals to spondylolisthesis (Mays 2006b; Merbs 2001). Steeply inclined sacral tables are risk factors for both spondylolysis and spondylolisthesis (Mays 2006b). Another anatomical variance that leads to spondylolisthesis risk includes extra lumbar or sacral vertebrae. Aebi (2012) found that lumbarization and sacralization led to both higher risks of spondylolysis and spondylolisthesis.

Looking at degenerative spondylolistheses, which are likely the most common type of vertebral slippage, Merbs (2002) studied spondylolisthesis in prehistoric Native American skeletal remains housed at the Smithsonian and compared the data to modern US patient data. In the 491 Native American individuals, Merbs found that spondylolisthesis, which increased with age, was found nearly always in conjunction with osteoarthritis, but osteoarthritis was not always found with spondylolisthesis. Spondylolisthesis was most frequently found in individuals over forty years of age, and the lumbosacral bones were most often affected. He found that these patterns have not changed over time and thus may reflect biological pressures.

Although males, it seems, have greater rates of spondylolysis, Merbs (2002) found that females have greater rates of spondylolisthesis. Merbs noted that the higher rates of spondylolisthesis in females compared to males in both prehistoric and modern samples were likely attributable to pregnancy's effect on lumbosacral morphology. Interestingly, Kalichman et al. (2009) found the same sex difference in data from the Framingham Heart Study. They suggested that females have more spondylolisthesis as a result of hormones and posture. Sex differences were reversed in isthmic spondylolisthesis, which makes sense if isthmic spondylolisthesis—unlike degenerative spondylolisthesis—relates to spondylolysis since spondylolysis is usually more frequent in males and was found to be more frequent in males than in females in the Framingham Heart Study.

SPONDYLOLISTHESIS AND PAIN. Spondylolisthesis patterns have remained fairly consistent over time, but whether spondylolisthesis causes pain is debatable. Once again, depending on sports literature is problematic since the individuals in the case studies or even in larger sample studies tend to take part in the research because of their existing pains. Reitman et al. (2002), for example, found that a sixteen-year-old gymnastic female experienced thigh pain related to spondylolisthesis, but Halpin (2012) noted that spondylolisthesis can be asymptomatic or symptomatic. Studies of skeletal remains, of course, cannot help us determine whether morphological traits lead to pain. Thus researchers are dependent on large-scale population studies, such as the Framingham Heart Study. Sonne-Holm et al. (2007) did not find pain related to spondylolysis or spondylolisthesis.

TREATMENT

Doctors seeing the physical traits of Schmorl's nodes, spondylolysis, or spondylolisthesis may decide to treat patients through exercise, medication, and surgery, even though physical traits found on the skeleton may not be correlated with pain. Some studies have found that exercise has no effect on back pain (e.g., Indahl 2004), and pain medications, such as non-steroidal anti-inflammatory drugs (NSAIDS) and cortisone, have usually been found to be ineffective (e.g., Ridgewell 2003). Due to the ineffectiveness of conservative therapies to treat back pain, patients have sought out alternatives including chiropractic care and surgery. McMorland and Suter (2000) found in individuals with chronic lower back pain that only 20% experienced relief through chiropractic care. Back surgeries have increased by 220% between 1990 and 2000 in the United States (Chan and Peng 2011). The overall success rate of back surgery in the United States is about 50% (Chan and Peng 2011). In Sweden, data from the National Registry provides evidence that two years after surgery, a quarter of patients who had surgery for hernias and a third of patients who had surgery for slippage experienced pain again (Strömqvist 2002). Furthermore, Slipman et al. (2002) found that in a sample of 267 individuals operated on, 61% experienced pain within six months of their surgery. Masharawi et al. (2008) urged complete understanding of normal vertebral anatomy prior to conducting surgery; for example, using the Hamann-Todd Skeletal Collection, they learned that wedged vertebral bodies are normal and not necessarily indicative of degeneration. Sartoris et al. (1985) employed intact cadavers to determine normal aging changes in vertebrae versus inflammation in order to help prevent unnecessary medical treatments.

CONCLUSIONS

Back pain is a common ailment and may have a long history; tattoos on Otzi, the 5,300-year-old ice mummy, have been found to correspond to locations of frequent back pain in modern populations (Indahl 2004). Due to the prevalence of back pain, clinicians and anthropologists have been searching for anatomical factors that can be altered to reduce the pain. Hernias, such as Schmorl's nodes, vertebral slippage known as spondylolisthesis, and vertebral stress fractures called spondylolysis have all been found in modern humans and in skeletal collections. The patterns of affliction seem fairly consistent, with the exception that young athletes tend to have more of these traits than their non-athletic counterparts, and individuals with high BMIs tend to have more hernias and fourth-lumbar spondylolysis. The continuity of traits has led anthropologists and other researchers to propose risk factors that include bipedality, anatomical variation, and aging. Activities, too, have been suggested to cause some of these vertebral pathologies, but large database studies fail to find normal activities linked to these traits, and familial patterns suggest that some genetic factors are at play. In both skeletal and patient samples, anatomical variations that are likely rooted in the genes do seem to predispose individuals to fractures, hernias, and slippages. Yet all of these patterns are not of consequence to quality of life unless they are related to pain. Ties between these traits and pain are precarious; just because a trait is visible on a bone or X-ray does not mean that the trait is correlated to pain. Even without knowing that the traits cause pain, doctors and chiropractors have opted to treat the pain by fusing bones, pulling backbones, or manipulating disks manually. The results of these measures have been less than stellar. Ehrlich (2003) has suggested that back pain is more psychological than physical; for example, job dissatisfaction correlates with back pain. Fibromyalgia, which is a non-specific pain **syndrome**, is also found in back pain patients (Ehrlich 2003). Colella (2003) lists modifiable risk factors of back pain that include obesity, smoking, and depression. Ehrlich (2003) and Indahl (2004) both recommend cognitive therapy prior to treating patients through more drastic measures.

CHAPTER 6

Arthritis

ACCORDING TO THE ARTHRITIS FOUNDATION (2013), "ARTHRITIS IS A COMplex family of musculoskeletal disorders consisting of more than 100 different diseases or conditions that destroy joints, bones, muscles, cartilage, and other connective tissues, hampering or halting physical movement." The most widely discussed types of arthritis are ankylosing spondylitis, rheumatoid arthritis, and psoriatic arthritis, which are erosive forms of arthritis, and osteoarthritis. Secondary arthritis as a result of an injury or overuse is usually categorized under osteoarthritis.

EROSIVE ARTHRITIS

Anthropologists sometimes refer to non-osteoarthritis forms of joint disease as erosive arthritis. Erosive arthritis includes ankylosing spondylitis, rheumatoid arthritis, and psoriatic arthritis. These forms of arthritis seem to be caused by genetic variance. For the most part, ankylosing spondylitis, rheumatoid arthritis, and psoriatic arthritis are treated as rarities in the bioarchaeological record (e.g., Bass et al. 1974; Inoue et al. 1999; Kilgore 1989; Martínez-Lavin et al. 1995; Rogers et al. 1991; Zias and Mitchell 1996). Part of the reason for the lack of erosive arthritis studies in bioarchaeology is the difficulty in diagnosing the disease from skeletal material (see Miller et al. 1996; Rothschild et al. 1990). Even in large skeletal sample studies, such as research on the Hamann-Todd Collection, erosive arthritis is relatively rare and occurs in less than 3% of the populations examined (Rothschild and Woods 1991a). Furthermore, there is no evidence that erosive arthritis frequency or severity has changed over time. Thus, erosive arthritis types will be described briefly, but the bulk of this chapter revolves around osteoarthritis.

Ankylosing spondylitis is a chronic inflammatory disease that most severely affects the vertebral column and sacrum. It causes fusion of joints through the ossification of fibrous ligaments and thereby results in stiffness and pain. It can also make breathing difficult if it affects the thoracic region. According to the NIH (2013), ankylosing spondylitis occurs in about 3.5 to 13 individuals per one thousand people, and it affects males more often than females. Symptoms appear during adolescence or early adulthood. The cause of ankylosing spondylitis appears to be genetic, but its etiology may be multifactorial.

Rheumatoid arthritis, according to the Arthritis Foundation (2013), is a chronic autoimmune disease that results in inflammatory arthritis. Rheumatoid arthritis may cause damage to joints, cartilage, tendons, and ligaments. In severe cases, the ends of bones may erode. Some afflicted individuals will develop rheumatoid nodules. These nodules are lumps that form under the skin and over bony areas, such as fingers, elbows, and heels. Symptoms change often; sometimes joints feel warm, and at other times there may be pain and swelling around the joints. Systemic symptoms include fatigue, loss of appetite, fever, and anemia. Rheumatoid arthritis etiology is not fully understood, but the cause of the immune system's attack on its own body is fluid buildup in the joints and systemic inflammation. Almost 1% of people in the United States have rheumatoid arthritis, and it occurs nearly three times more frequently in females than males. Rheumatoid arthritis usually starts between the ages of thirty and sixty in females, and when it does occur in males the onset is later (Arthritis Foundation 2013). Examining the Hamann-Todd skeletal collection, Rothschild and Woods (1991b) found that rheumatoid arthritis occurred in 4.4% of females and in 0.7% of males.

Psoriatic arthritis is a joint problem that occurs with the skin condition psoriasis; both psoriasis and psoriatic arthritis are autoimmune diseases. Usually psoriasis occurs prior to the onset of the psoriatic arthritis. Psoriatic arthritis often only affects a few joints and has mild symptoms (NIH 2013). However, according to the Arthritis Foundation (2013), the symptoms, which include tender, painful, and swollen joints (especially in fingers and toes), reduced range of motion, and morning stiffness, can develop slowly or be sudden, and the symptoms can be severe. The cause of psoriasis and psoriatic arthritis is unknown, but a genetic etiology is assumed. Psoriasis affects about 2% of the population, but most individuals with psoriasis will not get arthritis from it.

The erosive arthritis conditions are fairly rare, most likely determined by genes, and not well documented in the archaeological record. Osteoarthritis is very different; it is well studied within large samples for both past and present populations. Many of the physical traits used to diagnose osteoarthritis in the skeletal materials are the same or similar to traits examined in X-rays and MRIs in living peoples. Furthermore, osteoarthritis etiology seems to be multifactorial, with both hereditary factors and environmental factors playing an important factor in its expression. Thus researchers have examined patterns to understand lifestyles of the past and present.

OSTEOARTHRITIS: DIAGNOSIS AND RISK FACTORS

Osteoarthritis (abbreviated as OA) is the most common skeletal pathology; and other than dental disease, osteoarthritis is the most ubiquitous pathology in the skeletal record (Weiss and Jurmain 2007). In the United States today, osteoarthritis disables about one in ten individuals sixty years of age or older (Bajwa et al. 2013). Osteoarthritis is not reversible and it worsens over time, but there are ways to slow the pain and progression, like moving to an arid environment and losing weight (Bliddal and Christensen 2006).

The term *osteoarthritis* is in itself a contentious issue; in North America, *osteoarthritis* is the preferred term for the gradual breakdown of cartilage that resides between adjoining bones of a joint so that the articular surfaces come into direct contact with one another. However, some osteologists prefer the term *osteoarthrosis* or *degenerative joint disease* since the suffix *-itis* implies a disease of an inflammatory nature. Although researchers previously thought that osteoarthritis did not involve inflammation, recently inflammation has been found to be a fundamental aspect of osteoarthritis (Punzi et al. 2005). Thus the term *osteoarthritis* seems more appropriate now than it was previously thought to be. In this chapter, the term *osteoarthritis* will be used.

Diagnosing Osteoarthritis

Osteoarthritis is a progressive degenerative disease that changes the morphology, composition, and mechanical properties of joints. The symptoms of osteoarthritis include joint pain and stiffening, but osteoarthritis can also be asymptomatic (Dillon et al. 2007). Clinical characteristics of osteoarthritis include changes in the space between connecting bones due to the loss of cartilage, cartilage hardening and thickening, and tears in cartilage tissues. Physical attributes that both anthropologists and clinicians look for in diagnosing osteoarthritis include **lipping**, which is the formation of bony **spicules** or osteophytes around or near the margins of the joint; **eburnation**, which is a degeneration process that results in a shiny polished area of bone where there has been bone-on-bone rubbing from lack of cartilage; and, porosity, which is the formation of small holes that are on the joint. Figures 6.1 through 6.3 display various joints with the three traits (osteophytes, eburnation, and porosity) anthropologists use to diagnose osteoarthritis. Figure 6.4 displays fusion of vertebrae resulting from osteoarthritis, but in addition, a **syndesmophyte** is present. Syndesmophytes are bony growths attached to ligaments.

In clinical studies, osteoarthritis is most often diagnosed through the use of X-rays or MRIs. A commonly used scale, which is used in large databases such as the NIH Osteoarthritis Initiative, is the Kellgren and Lawrence osteoarthritis scale. The Kellgren and Lawrence scale is obtained through examination of radiographs and MRIs and is utilized for absence or presence of osteoarthritis

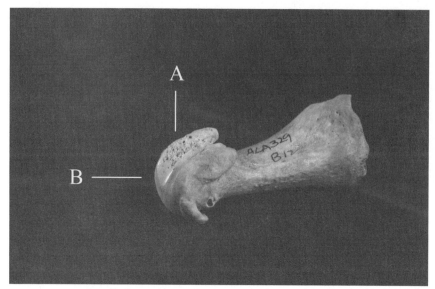

Figure 6.1. Metatarsal OA. An example of OA on the foot bone of a thirty-nine- to forty-four-year-old female. This bone displays porosity (A) and eburnation (B). Photograph by Daniel Salcedo.

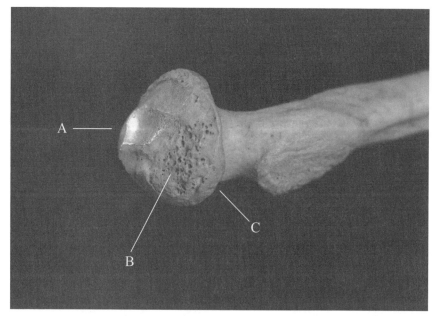

Figure 6.2. Radial OA. The radial head from this thirty-nine- to fifty-year-old female is hardly recognizable due to the eburnation (A), porosity (B), and lipping (C) around the joint surface. Photograph by Daniel Salcedo.

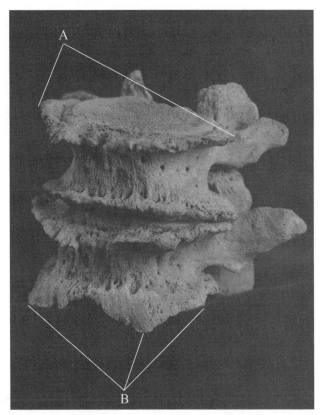

Figure 6.3. Vertebral OA. These lumbar vertebrae from a forty-one- to fifty-year-old male are a prime example of osteophyte lipping from degenerative bone disease or OA. Although the lipping is on all vertebral body rims, it is most evident on the superior rim of the upper vertebra (A) and the inferior border of the lower vertebra (B). Photograph by Daniel Salcedo.

as well as determining its severity. The Kellgren and Lawrence scale is based on a zero-to-four scale and mainly utilizes evidence of joint space narrowing and osteophytes.

Bioarchaeologists do not usually need to rely on imaging techniques and score osteoarthritis using macroscopic examination of joints. Even though there are variations in scoring methods, scales from zero to three or four are common. For example, Weiss (2006) utilized a zero-to-four scale for vertebral osteoarthritis that ranged from zero, for no evidence of osteoarthritis, to four, which was complete fusion of joints. For the limb joints, the scale was modified from zero to three. In both incidences, osteophytes, eburnation, and porosity were considered. Osteophytes thus are used both in clinical and anthropological studies of osteoarthritis, which make it a good trait for

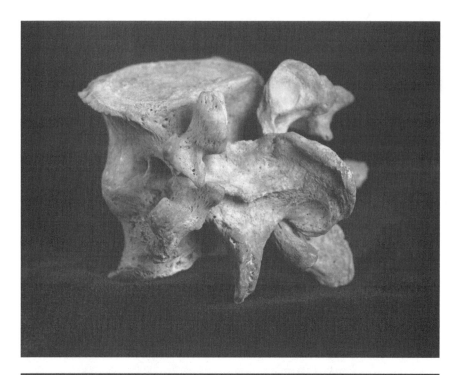

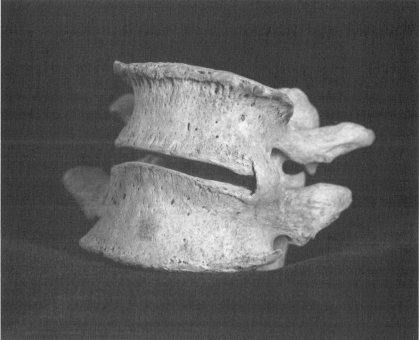

Figure 6.4. Vertebral fusion. Osteoarthritis sometimes results in vertebral fusion (top), which is displayed in a thirty-one- to forty-year-old male, coupled with a syndesmophyte (bottom). Photographs by Daniel Salcedo.

comparative purposes. Still, not all researchers are convinced that osteo-phytes are good indicators of osteoarthritis.

Although many studies include all three (eburnation, osteophytes, and porosity) characteristics to determine osteoarthritis presence and severity, research on osteophytes and porosity has led to questions about whether these traits are actually a result of osteoarthritis. As early as the 1970s, osteophytes had been called into question in regard to osteoarthritis diagnostics (Hernborg and Nilsson 1977; Jurmain 1977). Since then, many researchers, such as Moscowitz (1993) and Rogers et al. (1997), have reported that osteophytes seem to be age related and not pathological. In a study of Eskimos, Pecos Pueblo Amerindians, California Amerinds, and the Terry Collection, Jurmain (1991) ascertained that osteophytes have a stronger correlation with age than eburnation does. Woods (1995) and Rothschild (1997) suggested that porosity is a natural occurrence on bones and is not related to osteoarthritis. Schultz (1998) examined hip osteoarthritis in the Terry Collection and Hamann-Todd Collection and found that porosity was a poor indicator for osteoarthritis. Nevertheless, these traits are still used in both clinical and anthropological research in part because eburnation (which everyone agrees relates to osteoarthritis) is the least common of the three traits, and thus in many skeletal collections eburnation does not occur with enough frequency to determine patterns.

One way to determine whether the physical traits that occur on bones are really indicative of osteoarthritis is to see if they correspond to symptoms in clinical studies. Nguyen et al. (2011) found that osteoarthritis diagnosed with X-rays did not correlate with disability or pain. Hannan et al. (2000) examined a sample of nearly seven thousand individuals and found that about one thousand reported knee pain, but only 15% of these individuals had X-ray evidence of osteoarthritis. Furthermore, out of the 319 individuals with physical traits, such as cartilage narrowing and osteophytes, less than half (47%) reported that they were in pain. However, other researchers have found that pain corresponds to the physical traits seen in X-rays and MRIs. For example, Lethbridge-Çejku et al. (1995) used the Baltimore Longitudinal Study to examine 452 males and 223 females for osteoarthritic pain and symptoms, and they found that the 23% of people who reported pain were more likely to have X-ray evidence of osteoarthritis than those without pain. Duncan et al. (2006) suggested that some of the research that does not show that physical traits correspond to pain have taken X-rays from too few angles and that utilizing more images will reveal more symptoms. For now, osteophytes and eburnation are used in both clinical and skeletal studies; in clinical studies, traits also include cartilaginous changes, such as thickening and hardening, tears, and narrowing spaces between the joints. Porosity is the trait with the least amount of support for continuous use in diagnoses.

Osteoarthritis Risk Factors

MECHANICAL STRESSES. There are two main etiological explanations for osteoarthritis: localized deterioration of joints as a result of mechanical stresses or systemic deterioration of joints as a result of biological factors. Those who think that osteoarthritis is heavily influenced by mechanical stresses tend to be those who link osteoarthritis with activities. As early as the 1960s, anthropologists had reconstructed activity patterns using osteoarthritis (e.g., Angel 1966; Ortner 1968; Wells 1962). Since then, anthropologists have mostly attributed long-term mechanical stress as the cause of osteoarthritis, and thus they have used osteoarthritis as an activity indicator. However, most anthropologists have acknowledged the complex etiology of osteoarthritis formation, which includes genetic and environmental influences (see Bridges 1992; Rogers et al. 2004; Weiss and Jurmain 2007), and in the late 1990s there was a reduction in studies that used only osteoarthritis to reconstruct activity patterns, even though osteoarthritis was and still is frequently combined with other activity indicators for lifestyle reconstruction (e.g., Cope et al. 2005).

Repetitive tasks, anthropologists have argued, cause wear and tear on joints, and thus the pattern of affliction can help to reconstruct activity patterns. An example of using osteoarthritis to reconstruct activity patterns comes from work on eleventh- to thirteenth-century Colombians who had high levels of vertebral osteoarthritis, which Rojas-Sepúlveda et al. (2008) attributed to carrying heavy loads. The ability to determine the activity was aided by cultural continuity; populations in the region still engage in carrying heavy loads on their heads, so it seems likely that this may have occurred in the past. Lieverse et al. (2007) looked at populations separated by a seven-hundred-year gap in Siberia; the early population dated between 6800 and 4900 BC seemed to have more knee osteoarthritis and more pronounced sex differences than the later population dated between 4200 and 1000 BC. Lieverse et al. concluded that these differences imply a change in lifestyle after the gap, which may have involved an intensification of local resource use that reduced mobility. Cope et al. (2005) examined hand bones from the United Arab Emirates dating between 2500 and 400 BC and attributed high levels of hand osteoarthritis to the production of fish nets.

Activity studies linking osteoarthritis to mechanical stresses can also be found in clinical research. For example, Bernard et al. (2010) found, in a sample of 3,435, that hand osteoarthritis increased in those who engaged in jolting sports. Yet a study of rock climbers by Sylvester et al. (2006) found no increase in osteoarthritis in relation to the activity. Some of the best studies on activity patterns and osteoarthritis come from research on farmers. Farmers have higher rates of hip osteoarthritis than the general population, but the knee osteoarthritis rate results are mixed. Sandmark et al. (2000)

found that knee osteoarthritis was higher in both sexes in Swedish farmers compared to non-farmers; however, Holmberg et al. (2004) reported that the increase in knee osteoarthritis in Swedish farmers was present only in females. Anderson and Felson (1988) also suggested that knee-bending occupations were correlated with knee osteoarthritis. Agricultural workers start their work early in life, and this may be one of the reasons that they have an increased risk of osteoarthritis (Cooper et al. 1996; Rossignol 2004). Another possibility is that agricultural workers face more extreme stresses than the average person; extreme loads have been known to override genetic propensities (e.g., Manek et al. 2003). Finally, arthritis can be caused by injuries, and farming injury rates are high. Some of the correlations with arthritis and sports (e.g., Schmitt et al. 2004; Shepard et al. 2003) may also relate to secondary arthritis that is the result of an injury. Yet Chen et al. (2010) using the Hamann-Todd skeletal collection discovered that osteoarthritis of the shoulder joint is just as likely in individuals who had an injury as in individuals who did not experience a shoulder injury.

GENETICS. Those who focus on systemic etiology of osteoarthritis suggest that evolution and genes are of utmost importance (e.g., Hunter and Eckstein 2009; Jonsson et al. 2003; Rogers et al. 2004; Sigurdsson et al. 2008). These researchers argued that the universality of osteoarthritis suggests that the root etiology must span cultural and temporal explanations. Twin studies—some with good controls for age, BMI, and bone mass—that examine differences between monozygotic and dizygotic twins have reported that heritability of osteoarthritis is around 0.5, which means that half of the variance in osteoarthritis can be explained by genes. For example, Sambrook et al. (1999) used a sample of over three hundred twins between the ages of thirty-one and eighty years to study spinal osteoarthritis. In their study, the British twins revealed a heritability of 74% for spinal osteoarthritis. Page et al. (2003) examined 6,419 male veteran twins to look at the genetics of hip osteoarthritis; they found that additive genetics explained 61% of the variance of hip osteoarthritis seen in X-rays, while 39% was due to unique environmental factors.

Even though twin studies are ideal for genetic research, other familial studies can also provide information on heritability. For example, Lanyon et al. (2000) looked at siblings and risk of hip osteoarthritis in a large clinical study; he found that if an individual had a sibling with hip osteoarthritis, then the individual had a six times greater risk of hip osteoarthritis than the general population. Jonsson et al. (2003) used a large sample of nearly three thousand Icelanders to examine the heritability of hand osteoarthritis; they found that even beyond the nuclear family, family history of osteoarthritis increased the risk of osteoarthritis. Plus, different joints have different rates of heritability, which suggests to Weiss and Jurmain (2007) that not all joints are equally useful or useless for activity pattern reconstructions. For example, the spine and hip may have the highest heritability rates, with 60% to 70% of the variance being explained by genetics (Sambrook et al. 1999; Spector and MacGregor 2004), whereas the hand has a lower heritability rate of around 0.40, or 40% (Spector et al. 1996). The knee joint heritability rates vary; for example, in Spector et al. (1996), 0.40 is the heritability for the knee, whereas in Zhai et al. (2007), it may be as high as 0.60 to 0.70. Even within a joint, different parts of the joint are affected differently by environmental and genetic forces (Valdes et al. 2004; Zhai et al. 2007).

Genetic researchers have also discovered as many as nine polymorphic loci that relate to osteoarthritis (e.g., Bergink et al. 2003; Spector and MacGregor 2004; Seki et al. 2005). Even though this may seem like the nail in the coffin for activity-related causes of osteoarthritis, the etiology of osteoarthritis appears complex and involves **pleiotropy**, biological sex influences, and population variation on both genetic and environmental levels. For example, the genetic research has found alleles for severity risk, but not risk of absence or presence (see references within Weiss and Jurmain 2007). And females have higher heritability rates than males, which may be tied to estrogen receptor differences between the sexes (Bergink et al. 2003; Spector and MacGregor 2004; Wilson et al. 1990). Finally, there seem to be exceptions to the high heritability of osteoarthritis; extreme loads seem to affect the onset and severity of osteoarthritis regardless of genes (Manek et al. 2003). Thus, it may be that heritability is inflated in twin studies; Weiss and Jurmain (2007) suggested that the environment may play a greater influence on past populations when the stresses of life were more extreme and activities were started early in life. Regardless of whether one focuses on mechanical or genetic causes of osteoarthritis, there are some patterns that appear repeatedly and cannot be explained by culture.

AGE. The best overall correlation with osteoarthritis is age; older individuals have more osteoarthritis than do younger individuals. And, because of the increase in the population of older individuals, osteoarthritis will become a bigger concern for years to come. Researchers have often attributed this increase in osteoarthritis with age to an effect of mechanical stresses over time (Kahl and Smith 2000; Merbs 2001; Waldron 1997). However, excessive stress early in life may lead to early-onset osteoarthritis, which is what is found in bioarchaeological samples, but it may not relate to late-onset osteoarthritis that is seen in modern samples. Van de Westhuizen and Mennen (2010), for instance, stated that hand osteoarthritis starts late in life, but Cope et al. (2005) found early-onset hand osteoarthritis in her Bronze Age sample from the Middle East. Lee and Riew (2009) noted that cervical facet arthritis occurred in 30% of individuals over the age of seventy in the Hamann-Todd Collection, whereas Bridges (1992) examined Alabama Native American remains and found that cervical osteoarthritis was frequent in fairly young individuals, which she attributed to carrying objects on the head. Jurmain (1977) found that proto-historic Eskimos had earlier-onset knee and hip osteoarthritis compared to the Terry Skeletal Collection from the 1900s. And Merbs (1983) also found early onset and high rates of osteoarthritis in cold climate populations; he attributed these findings to hunting and kayaking, but cold climates, especially if they are high in humidity, can increase the risk of osteoarthritis (Kalichman et al. 2011).

Studies on the twentieth-century collections, such as the Terry Skeletal Collection and the Hamann-Todd Collection, have found that age is an important etiological consideration, especially in regard to the formation of shoulder and hip osteoarthritis. Weiss and Jurmain (2007) have reviewed data from prehistoric, proto-historic, and historic collections to assess which joints are most affected by age and which are more likely influenced by localized factors, such as mechanical stresses; their conclusion was that the elbow is least affected by age and thus may prove most useful in understanding activity patterns. Conversely, hip, knee, spine, and shoulder osteoarthritis are greatly influenced by age. Clinical studies on osteoarthritis have reported that at sixty-five years or older, nearly every other person has osteoarthritis.

ETHNIC AND SEX DIFFERENCES. Osteoarthritis exists worldwide; no population seems spared, although variation exists (van Saase et al. 1989). Some of the variation may be a result of activity differences; Zhang et al. (2004), for example, suggested that Chinese elderly may have higher rates of tibiofemoral (knee joint) osteoarthritis as a result of prolonged squatting. Nevitt et al. (2002), looking at a sample of around 1,500, found that Beijing elderly were less likely to have hip osteoarthritis, which the authors attribute to a combination of more walking in the Chinese population and genetic differences. Using a sample of nearly five hundred skeletal remains from individuals who died between 1913 and 1933, Rothschild and Woods (1991b) found that blacks had nearly twice the rate of osteoarthritis than whites did. In 1977, Jurmain found that in the Terry Collection, blacks had greater rates of knee osteoarthritis than whites. And Anderson and Felson (1988) found that when using the large CDC-based NHANES database, even with controls for age and weight, black females were more at risk for knee osteoarthritis than were white females. Allen et al. (2010) found that African-Americans expressed higher rates of pain than European-Americans in relation to osteoarthritis, which they attributed to higher BMIs and depression. Yet Ang et al. (2003) suggested that this population difference in pain is actually a result of greater osteoarthritis severity. Ang et al. stated that pain complaints from black and white male veterans are equal when osteoarthritis severity is controlled for. Another perspective is that population differences are hard to replicate and are likely a result of **interob-server error rates** (van Saase et al. 1989).

Sex differences are more consistent. In the archaeological record, sex differences in osteoarthritis frequency and severity have been attributed to divisions of labor. For example, Derevenski (2000), looking at populations in the United Kingdom dating between the sixteenth and nineteenth centuries, found that females had higher rates of vertebral osteoarthritis than males, which Derevenski attributed to female use of creel baskets. Sex differences, however, may be caused by factors other than activity patterns. Weiss and Jurmain (2007) and Waldron (1997) reviewed evidence that body size and hormones may affect osteoarthritis expression. Sex differences in the pelvis may help to explain the high rates of lower limb osteoarthritis in females, but Wilson et al. (1990) have linked these differences to genes.

In modern clinical studies, researchers tend to find that females have higher frequencies of osteoarthritis and experience more osteoarthritis pain than males (Hunter et al. 2005). The main sex differences occur in the knee and hip; females have more hip and knee osteoarthritis than males (van Saase et al. 1989). Research on other ailments suggests that female susceptibility is also genetic. For example, females tend to have higher rates of autoimmune diseases, and experimental research on rodents has provided evidence that females are more sensitive to pain (Aloisi et al. 1994). Females have higher rates of heritability than males in osteoarthritis, which could be caused by estrogen receptors (Bergink et al. 2003; Spector and MacGregor 2004).

ANATOMY EFFECTS. Although mechanical stress arguments of osteoarthritis etiology are usually focused on repetitive stresses caused by activities, such as occupations or labor, anatomical variation can also influence the stress experienced by specific joints. Bipedality and spinal curvature seem to increase the compressive forces that lead to spinal osteoarthritis (Bridges 1994; Kilgore 1990; Merbs 1983). And long tibiae that place the knee higher can increase the risk of knee osteoarthritis in females (Hunter et al. 2005). A long tibia may cause the knee to be less stable, increase torsion forces when walking, and thereby cause greater stress. Shallow acetabula (hip sockets) have also been found to increase osteoarthritis compared to deeper acetabula. Using a sample of 835 men and women, Reijman et al. (2005) found that individuals with **acetabular dysplasia**, which is measured by the depth of the acetabulum at its center and by the angle of the center-edge of the femoral head to the lip of the acetabulum, had a 4.3 times greater risk of hip osteoarthritis than individuals with normal acetabula. Hip **dysplasia** is a result of an abnormal femoral head, such as the congenitally deformed femoral head displayed in figure 6.5, and a shallow acetabular. And sometimes anatomical differences related to anatomy can cause population differences in osteoarthritis severity; for example, in the

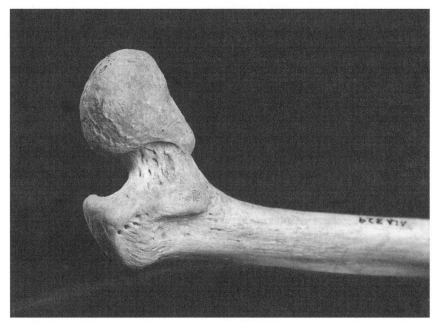

Figure 6.5. Congenitally deformed femoral head. This femoral head from an adult female between the ages of twenty-one and thirty years old. Photograph by Daniel Salcedo.

United States, black females have longer tibiae and more knee osteoarthritis than white females (Anderson and Felson 1988). Also, it appears that Chinese individuals have higher rates of knee osteoarthritis (even though they tend to have shorter tibiae) than Caucasians, which has been linked to higher rates of knee misalignment in Chinese than in Caucasians (Felson et al. 2002).

BMI EFFECTS. Although in past populations body weight was not likely a factor in the onset of osteoarthritis, the increase in body weights and in BMI has had a tremendous effect on osteoarthritis in the recent past and in the present. An increase in BMI and waist circumference has been linked to an increase in osteoarthritis in multiple studies. Dumond et al. (2003) and Manek et al. (2003) suggested that the increase in BMI increases osteoarthritis risk and severity because the body is carrying more weight, and therefore this increases the mechanical stresses, especially on weight-bearing joints, such as the knee and lower back. The evidence for the effect of excess weight on knee osteoarthritis is especially strong; it has been stated that obesity is the primary modifiable trait to reduce knee osteoarthritis (Lee et al. 2013). Even with controls for age and lifestyle, BMI still correlates with knee osteoarthritis whether it is measured on X-rays, through MRIs, or from self-reported pain surveys (Baum et al. 2013; Niu et al. 2009). Furthermore, even childhood weight can affect the risk of osteoarthritis later in life. MacFarlane et al. (2011), using a sample of 8,579, found that weight at eleven years of age could be used to predict knee osteoarthritis.

It appears that obesity affects females more than males (Peltonen et al. 2003). D'Arcy (2012) reported that 69% of knee replacements in middle-aged females are due to obesity. Black and Hispanic females are especially prone to knee osteoarthritis; they also have the greatest rates of obesity (Losina et al. 2011).

Even though the arguments for BMI's correlation with osteoarthritis have usually centered around mechanical explanations, recent research on non–weight-bearing joints may tell a different story. Gabay et al. (2008) found that hand osteoarthritis and weight were correlated. Research on biochemicals that are active during bone healing suggest that inflammation rather than mechanical stress is the source of the osteoarthritis and excess weight link (Grabiner 2004). For example, leptin changes in the inflammatory state and can cause osteophyte production. The excess leptin in obese people may translate into a higher rate of osteoarthritis (Dumond et al. 2003). Plus, chronic inflammation is found in obese individuals (Weisberg et al. 2003).

Another piece of the BMI, osteoarthritis, and pain puzzle involves activity. Moderate levels of activity seem to reduce osteoarthritis pain and slow its progression in clinical populations, but obese individuals are less likely to be active than normal-weight individuals (Deyle et al. 2000). The question of whether people with high BMIs have more severe osteoarthritis pain and features due to their body weight or due to their lack of activity has yet to be fully addressed.

OSTEOPOROSIS AND OSTEOARTHRITIS. Due to the formation of excess bone in relation to osteoarthritis, some researchers have suggested that osteoarthritis has been selected for as a protection against osteoporosis. For example, Yahata et al. (2002) used a sample of 567 females and examined bone mineral density and hand osteoarthritis; they found that osteoarthritic females had higher bone mineral densities than non-osteoarthritic females. Sowers et al. (1991) also found a positive correlation between bone mineral density and osteoarthritis in females. Marcelli et al. (1995) examined a sample of three hundred females over the age of seventy-four and found that osteoarthritic females had fewer fractures than individuals without osteoarthritis. Jiang et al. (2008) found that in post-menopausal females, osteoarthritic individuals were less likely to have Colles' fractures. And, using a prehistoric California Amerind sample, Weiss (2013) found that hand osteoarthritis was positively correlated with hand **robusticity** that was measured as bone shaft diameter. Not all studies have been able to replicate the finding of osteoarthritis protecting from osteoporosis. For example, Burger et al. (1996) examined a sample of 1,700 individuals and found that femoral neck bone mineral density increased with osteoarthritis, but that bone loss in other locations occurred as well. Furthermore, Hochberg et al. (1991) used a sample of 888 males and found that hand osteoarthritis correlated with a decrease in cortical bone area.

OSTEOPOROSIS AND OTHER DEGENERATIVE DISEASES

Two other degenerative disorders may have an inverse relationship to osteoporosis; diffuse idiopathic skeletal hyperostosis (DISH) and hyperostosis frontalis interna. Both of these diseases are thought to be related to hormones and are classified as metabolic diseases. They have similar traits to osteoarthritis; that is, excess bone formation is present in each. With DISH it is presented as calcification of ligaments, and in hyperostosis frontalis interna it is presented as bone growth inside the skull as illustrated in figure 6.6. Due to the excess bone, the universality of these pathologies, and the consistent patterns across populations, I suggest that these may have been selected by evolution to prevent osteoporosis; that is, it may be that the same genes that code for DISH and hyperostosis frontalis interna code for higher bone density at older ages.

DISH, which is also known as Forestier's disease, is expressed as an excess of calcification of the anterior longitudinal spinal ligament and can cause stiffness; however, the symptoms are usually mild (Jankauskas and Urbanavičius 2008; Kiss et al. 2002). DISH can spread to joints beyond the spine. When it is not restricted to the spine, it often goes misdiagnosed (Garg et al. 2008). DISH

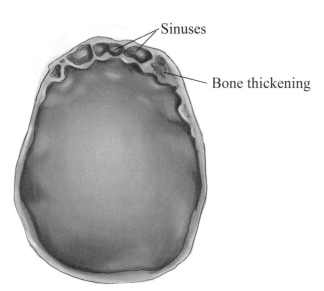

Figure 6.6. Hyperostosis frontalis interna. An illustration of a cranium cross-sectioned; the frontal bone is superior, which illustrates hyperostosis frontalis interna. Illustration by Vanessa Corrales.

increases with age, and it is more common in males than in females. Roths-child and Woods's (1991b) study of the Hamann-Todd Collection found that DISH was present in over a quarter of males over the age of sixty-four and only 5% of females in that same age range. Although DISH is not common in the bioarchaeological record, the afflicted individuals are most often males (e.g., Arriaza et al. 1993; Jankauskas and Urbanavičius 2008; Smith et al. 2013). The lack of DISH in the bioarchaeological record may be due to the late onset of DISH and the early age of death in many archaeological samples (Smith et al. 2013). Yet DISH has also been found in nonhuman animals, including dinosaurs, and is found globally, which suggests a biological etiology (Kiss et al. 2002). Even though genetics likely plays a role in DISH risk, individuals with higher BMIs and type II diabetes are at greater risk for DISH than the general population, which suggests that DISH may have increased over time and will likely increase in frequency in the future (Jankauskas and Urbanavi-cius 2008; Mathew et al. 2011). In the archaeological record, Jankauskas and Urbanavičius (2008) found no temporal changes, but they did find that DISH correlated with social class. The same link to social class is found in a six-teenth- to eighteenth-century Korean sample (Kim et al. 2012). In both cases, the elite likely had high calorie diets.

Hyperostosis frontalis interna, which correlates with DISH and is sometimes assumed to have a similar etiology, is more frequent in whites than in blacks and is more frequent in females than in males (Wilczak and Mulhern 2012). Hyperostosis frontalis interna is characterized by bone deposition on the in-ner table of the frontal bone on both sides, but not the midline. Hyperostosis frontalis interna is common in elderly females and likely is linked to hormonal changes after menopause. Chudá and Dörnhöferová (2011) examined the rates of hyperostosis frontalis interna between seventeenth- and eighteenth-century Slovakia and present Slovakians; they found that the past population had far fewer cases of hyperostosis frontalis interna than the present population. May et al. (2011) also found an increase in hyperostosis frontalis interna; in the last one hundred years, the rate has increased by 2.5 times. Furthermore, hyperostosis frontalis interna appears to be more severe and starts earlier today than in the past. The change may relate to fertility changes that include earlier menarche, later age at first birth, lower parity, hormone replacement therapy, and obe-sity. As early as 1719, Morgani linked hyperostosis frontalis interna to obesity and virilism, which is when male traits are found in females. Although about 40% of post-menopausal females have hyperostosis frontalis interna nowadays, Rühli and Henneberg (2002) stated that this disease was rare in archaeological samples, with only fifty-eight cases ever discovered. Rühli and Henneberg also associated hyperostosis frontalis interna to hormones; specifically, they sug-gested that hormones, such as estrogen, that are found in leptins may be key to understanding the temporal increase in hyperostosis frontalis interna. On the other hand, Hershkovitz et al. (1999) cautioned against comparing X-ray data to bone data in cases of hyperostosis frontalis interna; they also suggested that hyperostosis frontalis interna should not necessarily be considered pathological. Many individuals experience no symptoms, whereas other symptoms include headaches and hirsutism (excess in body hair growth).

JOINT PAIN TREATMENT

Most people choose to treat their osteoarthritis pain with medicines: acetaminophens, aspirin, ibuprofen, and the prescription drug Naproxen are the top choices of pain relief (Merkle and MacDonald 2009). Nonetheless, the side effects of these drugs, especially when used over long periods of time, may lead some patients to try alternative medicines. Some of the alternative treatments include glucosamine (e.g., Sanders and Grundmann 2011), herbs (e.g., Lechner et al. 2011), acupuncture (Sanders and Grundmann 2011), ointments (e.g., Crosby 2009), and passion-fruit peel extract (Farid et al. 2010). In a study that looked at multiple alternative medicines, Sanders and Grundmann (2011) examined glucosamine, devil's claw, and acupuncture. **Glucosamine** is a natural compound that is found in healthy cartilage. Devil's claw is a plant that has been thought to be anti-inflammatory. Acupuncture is a method of treatment where pins or needles are placed in certain body points to relieve people of pain. All three of these methods, according to Sanders and Grundmann (2011), are thought to be mildly effective, but the authors could not account for the **placebo** effect since it was not a **double-blind study**. Reginster et al. (2012) in a review paper found that glucosamine worked for knee osteoarthritis pain, but not for spinal pain. Although double-blind studies have shown the efficacy of glucosamine, the over-the-counter versions of glucosamine may not be as effective as the pills utilized in the clinical studies. Most alternative medicines seem to work for a short period of time; the pain may be reduced for days or weeks. Crosby (2009) reported that multiple alternative treatments may decrease pain momentarily, but that the effect may also be driven by the placebo effect. Ragle and Sawitzke (2012) have called into question the validity of alternative medicines; they found that standardized testing is missing and that research has produced mixed results. Braces may be an effective way to control OA pain (Page et al. 2011), but Crosby (2009) found that taping was as effective as expensive braces. Additionally, injected solutions are a common way to treat osteoarthritis. For example, intra-articular hyaluronic acid and cortisone have been used to treat osteoarthritis pain, but Colen et al. (2012) found that there was little evidence that hyaluronic acid was better than placebo, and Crosby (2009) noted that the efficacy of cortisone was low to medium and relief lasted only a few weeks.

Exercise has proven to be effective for reducing osteoarthritis pain. For example, Paans et al. (2013) found in individuals with hip osteoarthritis that exercise had lowered pain scores and that they did better in walking tests. In another study, Hiyama et al. (2012) found that adding steps to other conservative pain treatments, such as moderate exercise and ice packs, improved patient knee osteoarthritis symptoms. Yan et al. (2013) found that tai chi has a positive effect on pain over a twelve-week course.

Eighty-five percent of knee and hip replacement surgeries are done to treat osteoarthritis pain and stiffness (Gidwani and Fairbank 2004; Williams 2003). Complications may include **hemorrhaging**, **deep vein thrombosis**, and infection (Williams 2003). Metal-on-metal hips may cause metal debris to enter the body, and this can cause serious systemic effects, such as deafness, blindness, and cognitive decline (Campbell and Estey 2013). Uneven leg lengths are also possible after hip surgery (Knutsson and Engberg 1999). Unnanuntana et al. (2010) cautioned surgeons against use of a single measurement to determine appropriate limb length symmetry; using the Hamann-Todd Collection, they found that a frequent measure utilized by surgeons, the measure from the tip of the greater trochanter to the center of the femoral head, is an inaccurate measurement to determine limb length. From the quarter of a million hip replacements each year in the United States, one in five will eventually undergo revision surgery for loosening of the replacement, infections, instability, and mechanical complications (Biau et al. 2012). And knee replacement failures occur in about 16% of patients (Cloke et al. 2008). A common complication from hip surgery includes heterotopic bone formation, which is the formation of bone in a location where bone should not exist. Heterotopic bone formation can cause stiffness, pain, and fusion of the joint. Neal et al. (2002) reported that four out of ten individuals with hip surgery may have some heterotopic bone formation, and nearly one in ten will have severe ankylosing. Although surgery may seem like a permanent solution, Knutsson and Engberg (1999) found that pain returned within half a year to nearly half of the patients they examined for hip replacement. Katz et al. (2013), using two self-reported pain surveys, also found that surgery for tears in the **meniscus** that are the result of knee osteoarthritis is not more effective than physical therapy one year after surgery.

CONCLUSIONS

The most common form of arthritis is osteoarthritis; this disease usually occurs in older individuals, but early onset may occur as a result of severe trauma or excessive activities. Although there is a strong genetic component to the expression of osteoarthritis, the environment may still play a role in its etiology. In past populations, it appears that osteoarthritis started earlier in life and was more likely associated with activity patterns. However, current developed nations' populations seem to have been influenced less by their activities and more by their weight and lifestyle. Due to an increasingly older population, osteoarthritis severity is likely to be a growing health issue. When coupling the aging population with obesity, osteoarthritic individuals will likely be at greater risk for disability. So far, research suggests that there are no cures for osteoarthritis, and the disease worsens over time. Although there are a multitude of studies on osteoarthritis treatment, many of them are not double-blinded. Double-blind studies, in which the patient and the examiner do not know who received treatment, are thought of as the gold standard in medical research. Perhaps the best method for reducing the speed of osteoarthritis progress and for reducing its impact on quality of life is to lose weight, if needed, and keep moving. Exercise may seem counterintuitive in regard to osteoarthritis treatment since some of the environmental evidence has implied that osteoarthritis is a result of mechanical stresses and repetitive activities, but halting exercise and movement actually increases the risk of becoming disabled by the disease once arthritis has started. I often tell students that osteoarthritis should not even be considered pathological since if one lives long enough, osteoarthritis is nearly inevitable. People all over the world have osteoarthritis, and even nonhuman primates, such as gorillas, chimps, and bonobos, get osteoarthritis (see Jurmain 1999). However, not everyone with the physical manifestations of osteoarthritis experiences pain, and the severity of pain varies greatly. Some of this variance is genetic, but the modifiable factors that influence osteoarthritis expression include exercise and body weight.

CHAPTER 7

Oral Health

According to the WHO (2013), the latter part of the twentieth century saw a transformation in oral health, but millions of people worldwide have been excluded from the transformation. The changes that have occurred affect both children and adults and have a wide range of effects on overall health. Regardless of the progress that has been made, dental diseases are the most common diseases in both past and present populations. Dental diseases can be divided into tooth wear (**attrition, erosion,** and **abrasion**), **caries** (cavities or tooth decay), **periodontal disease** (gum and bone disease), and tooth loss. However, these various forms of dental disease—and some other dental problems, such as **malocclusion**—are not independent of one another. For instance, periodontal disease often leads to tooth loss. And malocclusion can lead to an increase in tooth decay, whereas attrition may lead to a decrease in tooth decay.

TOOTH ANATOMY AND EXAMINATIONS

Tooth anatomy, which is illustrated in figure 7.1, includes **enamel** (the hard mineralized tissue on the outer layer of the tooth), **dentin** (a bone-like structure that makes up the majority of the crown), **pulp** (unmineralized connective tissue that is **vascularized** and has **lymph cells**), **cementum** (mineralized tissue that holds the tooth in the tooth socket), and the **periodontal** membrane (location for nerves and blood supply). The tooth root sits in the tooth socket, which is called the alveolus, in the **maxilla** (upper jawbone) or **mandible** (lower jawbone). Humans have two sets of teeth; **primary dentition** (also known as milk teeth or deciduous teeth), which are replaced from ages six to fourteen years, and **secondary dentition,** or permanent teeth. Humans have twenty primary teeth and thirty-two permanent teeth. The permanent teeth include a pair (upper and lower) of two central incisors, two lateral incisors, two canines, four premolars, and six molars; there are no premolars in milk teeth. The different teeth have different shapes and purposes. The incisors are

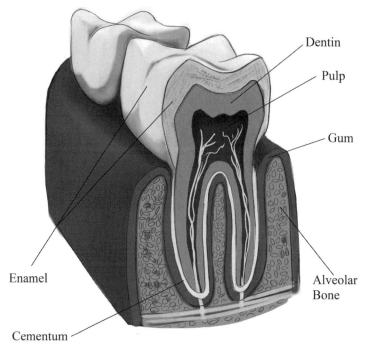

Figure 7.1. Tooth anatomy. Teeth, which are covered by mineralized enamel and anchored in bone by cementum, do not remodel like bone; the tooth is kept "alive" through the pulp, which is vascularized and has lymph cells. Illustration by Vanessa Corrales.

spade-like for biting into things, such as fruit; the canines are conical shapes for tearing; the premolars have two cusps for tearing; and the molars are **multicusps** for grinding. These **heterodonts** (different forms of teeth) reflect an omnivorous diet, a trait that humans share with other primates.

The mandible and maxilla will react like other bones and remodel in response to stresses. Teeth, too, remain dynamic after eruption. After root formation has occurred, secondary dentin forms. It forms slowly throughout life; the formation of secondary dentin reduces the pulp cavity, which makes it more difficult for microbial invasions. There is also tertiary dentin that is formed in reaction to trauma or microbial invasion, such as when a cavity occurs. Thus, although enamel does not remodel and repair, dentin reacts to prevent bacteria from entering the bloodstream (Charadram et al. 2013). However, if nutritional deficiencies occur, such as a lack of magnesium or zinc, secondary and tertiary dentin formation may be hindered; many bacteria utilize magnesium for growth and reproduction (Charadram et al. 2013). Furthermore, since enamel does not remodel, many dental pathologies remain visible on skeletal remains. Dentists and dental researchers may investigate dental pathologies through the use of X-rays and macroexamination of teeth.

Most pathology, such as tooth wear, caries, and malocclusion, are visible to the eye. Other methods dentists may use include stains and light to determine erosion or abrasion. Anthropologists tend to focus on macroexamination or scanning electron microscopes, which are microscopes that produce images by scanning the item with a focused beam of electrons, to document dental pathologies. Scanning electron microscopes can reveal microwear caused by foods rather than just the macrowear. Another way to investigate dental health is to use experimental models from animal teeth. And laser studies that map the **occlusal surface** (the biting surface) of teeth like a topographic map are also used in anthropological dental studies.

TOOTH WEAR: ATTRITION, ABRASION, AND EROSION

Tooth wear can lead to non-carious loss of tooth tissue. Tooth wear comes in three different forms: (1) attrition, which anthropologists define as tooth-on-tooth wear; (2) abrasion, which is defined as wear that comes from food or objects, such as toothbrushes; or (3) erosion, when **acidic** chemicals wear away the enamel on teeth. In each of these types of tooth wear, enamel is worn away exposing dentin. Dentin exposure can cause hypersensitivity and eventual exposure of the pulp, which can result in bacterial access to the bloodstream and systemic infections (Addy 2008). Figure 7.2 displays maxillary tooth abrasion.

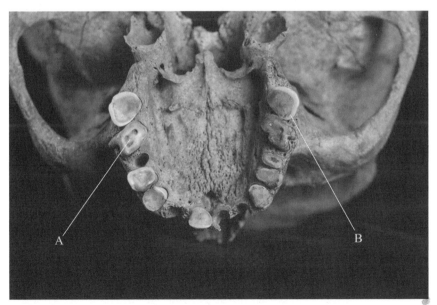

Figure 7.2. Tooth wear. The maxillary dentition on this adult male shows the extreme abrasion experienced by some hunter-gatherer-fisher populations. Although most of the teeth are worn down to pulp exposure (A), some of the teeth, such as the third molars, retain a ring of enamel (B). Photograph by Daniel Salcedo.

Tooth wear has some cross-cultural patterns; for example, the third molars, which are the last teeth to erupt, tend to be the least worn, and mandibular wear occurs more quickly than maxillary wear. Also, flat wear precedes cupped wear. Tough fibrous diets lead to flat wear, which may increase the chewing power. Inclined wear can be the result of malocclusion, coastal diets of shellfish, or an adaptation to increase chewing ability when teeth are severely worn (Watson 2008a). One of the main factors in tooth wear is age. Anthropologists have used occlusal wear as an age indicator, and age correlations with tooth wear have been found in populations as different as medieval Danes and Nigerians; but using tooth wear to age an individual can only be done using a within-population scale (e.g., Akapata 1975; Boldsen 2005; Mays 2002). Besides age, diet is the second most important factor in tooth wear (Boldsen 2005; Mays 2002). Population levels of wear are nearly always related to diet, whereas extraneous factors, such as tooth grinding, are revealed in interindividual variation.

When considering diet, anthropologists have recorded that prehistoric skeletal remains have greater occlusal wear on their teeth than more recent populations (Mays 2002). Bioarchaeologists have suggested that these wear changes are driven by environmental rather than genetic factors since the changes occurred in short time periods associated with subsistence shifts from hunting and gathering to agriculture. Generally, hunter-gatherers have more occlusal wear than agriculturalists. For example, Deter (2009), using a sample of over two thousand teeth from North American populations that included three Late Archaic hunter-gatherer populations and a late Anasazi/early Zuni agricultural population, found that the hunter-gatherers had higher rates of occlusal dentin exposure than the agriculturalists. However, there are exceptions; for example, Mahoney (2006) examined how the shift from hunting and gathering to agriculture affected dental wear in Israel and found an increase in pitting in the dentition of the agricultural population due to the use of stone-ground foods, which would have added an abrasive quality to foods.

Food processing seems to be the greatest cause of less tooth wear; changes in food processing can alter tooth wear even when diet has remained the same. For instance, Chattah and Smith (2006), who worked on remains from the southern Levant, argued for the importance of food processing as a factor of dental wear rather than diet itself. And Watson (2008a) examined samples from northwest Mexico during the early agricultural period and noted that in the sample of eighty-four skeletons from the early agricultural period, dental wear had changed even though there was no evidence for dietary changes. He suggested that the shift in greater molar occlusal wear related to an increase in the mechanical processing of food that increased dietary grit.

A minor factor in the reduction of tooth wear may be an increase in tools. Molnar (2008) examined tooth wear in a prehistoric Swedish sample and found that these Neolithic Swedes had striations that occurred in between teeth that were likely the result of pulling fibers between teeth either for pain relief or to prepare for baskets. The use of teeth to prepare fibers goes back nearly eight thousand years ago in Libya (Minozzi et al. 2003). Even present-day non-Western populations may use their teeth as tools; the Hadza of East Africa use their teeth in multiple ways (Berbesque et al. 2012).

Although tooth wear has generally decreased, two types of non-carious tooth traits have increased. Non-carious cervical lesions are rare in pre-industrial populations, but are frequent in modern samples (Benazzi et al. 2013). Non-carious cervical lesions are found in between teeth and at the junction between the crown and the root. Benazzi et al. (2013) proposes that they are a result of tensile forces placed on teeth for long periods of time. The lack of dental wear has lead to the retention of crown height, which increases the tension experienced by the cervical region, and that, coupled with a long life, may result in an increase in non-carious cervical wear.

Dentists emphasize the importance of good dental hygiene to reduce tooth decay, gum disease, and tooth loss, but researchers have also examined whether brushing and flossing may lead to abrasion. Brushing seems to do no harm to the hard dental enamel, but some very abrasive toothpaste may cause more tooth wear than others. Toothpastes that emphasize stain removal and whitening seem to be more abrasive than other toothpastes (Addy 2008). Toothbrushing, whether manual or electronic, seems to have only minor effects of abrasion and can only cause abrasive lesions when the teeth are already damaged (Wiegand et al. 2012). Toothbrushing can be harmful when it is done too soon after consumption of acidic drinks (Hemingway et al. 2006).

Although erosion has been found in *Homo habilis*, pre-contact New Zealand Maoris, and four-thousand-year-old skeletons from Brazil, erosion is rare in prehistoric dental remains (d'Incau et al. 2012). Erosion has increased over the last six decades (Wegehaupt et al. 2011). Part of that increase may be due to the aging population, since erosion is a normal part of aging (d'Incau et al. 2012; Nunn 1996). But it may also be a result of a dietary change. As early as the 1950s, erosion has been identified as a dental problem linked to acidities in sodas, foods, alcohol, and fruit juices (d'Incau et al. 2012; Nunn 1996; Wegehaupt et al. 2011). Research in the 1950s in the United Kingdom found that fluoride added to grapefruit decreased erosion, but too much fluoride is toxic and thus cannot be added to all acidic drinks (Wegehaupt et al. 2011). The earliest population studies on dental erosion and fruit exposure occurred in 1972 in Southern California; the study found that 18% of the sample had erosion (Nunn 1996).

This rate of erosion is fairly high, but in Roman/Anglo-Saxon remains, nearly 20% of individuals had erosion, which was likely linked to vomiting. Vomiting increases erosion due to gastric acids (d'Incau et al. 2012; Nunn 1996). In the 1970s, fad diets of grapefruit and other fruit combined with an increase in fruit juices for health led dental researchers to warn of the link between excessive exposure to citric acid and tooth erosion (Nunn 1996). Erosion can lead to pain through dentin exposure, and it is difficult to fix; but research on supplements, such as calcium and phosphate ions added to orange juice, suggest that the erosiveness of juices can be greatly reduced (Wegehaupt et al. 2011).

TOOTH DECAY

Tooth decay, also known as cavities and caries, is the most common dental disease in the world. In the 1970s, there was great debate over the cause of tooth decay; some researchers thought that diet played the key role in the formation of caries, while others emphasized the importance of bacteria. Scientists now know that tooth decay is the result of mineral-eroding bacterial discharge, which is usually a result of *Streptococcus mutans*. Caries (also known as tooth decay), which is illustrated in figure 7.3, is a disease that is characterized by

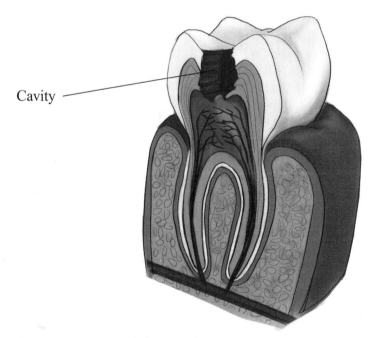

Cavity

Figure 7.3. Caries. Tooth decay results in the softening of enamel and eventually causes a hole (or cavity) in the tooth, which is called caries. When the hole reaches the pulp, this can let infectious agents into the bloodstream. Illustration by Vanessa Corrales.

demineralization of the dental hard tissue of enamel and dentin from the acid produced by bacteria (mainly *S. mutans*) that feed upon dietary carbohydrates (Watson 2008b). Caries etiology, although linked to *S. mutans*, is multifactorial; there are many causes of caries formation, including poor dental hygiene, a **cariogenic** diet, and genetic predisposition.

The bioarchaeological literature is rife with studies on caries rates. Caries are noted by the dental community through X-rays and examination of teeth; when a soft dark spot occurs on the tooth, this is usually diagnosed as caries. In skeletal remains, caries are usually seen as holes in the tooth; the holes or craters are visible to the eye. Since tooth loss was common in past populations and because of postmortem tooth loss, caries rates reported in the bioarchaeological literature differ slightly from clinical research. Still, trends are comparable. Anthropologists have reported that prehistoric caries rates increase in the New World with the adoption of agriculture, especially maize agriculture (e.g., Larsen 1995), but some studies fail to find this trend (e.g., Lanfranco and Eggers 2010). In Europe and Asia, agriculture adoption also is not always associated with an increase in caries. In Bulgaria, Keenleyside (2008) examined fifth- to second-century BC skeletal remains of 162 individuals and found high caries rates, which she suggested may be associated with a diet of soft foods and high carbohydrates. And Temple and Larsen (2007) found evidence for increased caries frequency in Japan that was coupled with the adoption of rice agriculture, but in Thailand and Cambodia rice adoption did not lead to an increase in caries (Halcrow et al. 2013). In the Old World some of the agricultural foods, such as millet and wheat, are not considered to be as cariogenic (likely to produce caries) as other agricultural foods. Rice, although it has been linked to an increase in caries, has been described as a low cariogenic food due to its low acidity and low sugar component.

Part of the increase in caries with the adoption of agriculture may be the result of a decrease in attrition or abrasion. Caries tend to be more frequent on cusped (fissure and pit) surfaces than on smooth surfaces, and wear changes cusped teeth to smooth teeth. For instance, Palubeckaitė-Miliauskienė and Jankauskas (2007) found that Napoleonic War soldiers from 1812 had more wear than German World War I soldiers from 1915 to 1917, but the German soldiers had more severe caries (although not more frequent caries). Yet in some cases there are exceptions to this inverse relationship. For example, Eshed et al. (2006) examined evidence of wear and caries in pre-agricultural populations and early agricultural populations of the Levant and discovered that occlusar wear was higher in the pre-agricultural population compared to the Neolithic agricultural population, but no differences in caries between the two populations were noted. In another study in the Old World, Arnold et al. (2007) examined Ukrainian samples from the eighth to tenth centuries

AD and found that the hunter-gatherers had lower wear rates than did the agriculturalists, but there were no differences in caries rates.

There are some prehistoric sites where specific foods that are sticky contributed to a high caries rate. Watson (2008b) examined caries at the La Playa site in northern Mexico and reported that non-agricultural Amerinds at the site had the same caries and tooth loss rates as the agriculturalists over a 1,500-year time span (2000 BC to AD 500), likely because the hunter-gatherer diet of wild carbohydrates included cacti, beans, and agave that are sticky and as cariogenic as maize. Nelson et al. (1999) reported high rates of caries in Iron Age (100 BC to AD 893) Omanian remains, which they associated with date consumption. Márquez-Grant (2009) found a high rate of caries as a result of sticky foods, such as figs, combined with cereals and high meat consumption in sixth- to second-century BC Spaniards. Meat is a low-abrasion food, and thus the high rate of terrestrial meat consumption coupled with figs may have led to an increase in caries even without cereals.

In the last forty years, caries rates have decreased drastically in some regions. Tooth decay is largely preventable. Sealants, some foods (such as xylitol), and fluoride prevent caries that are caused by *S. mutans*. From the 1970s to the 1990s, there has been a 42% decrease in caries in primary teeth in two- to eleven-year-olds in the United States (NIH 2013). The use of sealants has also decreased cavities (Beltrán-Aguilar et al. 2005). Dental sealants are highly effective and have decreased caries rates by 60%. However, only a third of six- to nineteen-year-olds have dental sealants, and these sealants are more common in whites than in minority groups. Nevertheless, in 1988 to 1994, sealant use was 20%, and it increased to 32% in 1999 to 2002. Brushing and flossing have further reduced the risk of severe dental problems.

The main source of the decrease in caries comes from fluoride use, especially when fluoride is added to tap water. Mass fluoridation started in the 1940s in the United States; fluoridation of tap water seems to be correlated with a 40% decrease in caries (Horowitz 2003; Lalumandier and Ayers 2000). Fluoride in tap water has especially reduced cavities in poor people in the United States (Jones and Worthington 1999). Fluoride in the water has been linked to a decrease in tooth loss as well (Neidell et al. 2010). Fluoride in toothpastes has also increased, especially in the developing world. Just in the last decade, fluoride in toothpaste in China has become commonplace (Liu et al. 2007). The addition of fluoride even offsets some of the effects of negative behaviors people engage in; for example, Armfield et al. (2013) found that although soft drink consumption increased 500% from 1947 to 1999, fluoride reduced the cavity risk, and thus a strong link between soft drink consumption and caries risk is no longer found in all locations. The use of fluoride, in water and toothpaste, however, has led to hypomineralization of enamel, which results in white spots on teeth. Excessive fluoride consumption can also lead to gum disease (Vandana and Reddy 2007). Other anti-cavity solutions may be less toxic, such as theobromine, which is found in chocolate and has been found to have enamel-strengthening components (Sadeghpour and Nakamoto 2011).

Although caries rates have decreased, there are still many individuals who have untreated caries. The CDC (2013) estimates that in six- to nineteen-year-olds, 15% have untreated caries, whereas nearly a quarter of twenty- to sixty-four-year-olds have untreated caries. And dental visits in adults seem to be stuck at around 60%. According to the CDC's NHANES data, among whites an increase in income and education has led to an increase in caries among twenty- to sixty-four-year-olds, whereas lower-income minority groups have greater caries rates than their higher-earning counterparts. Some of the reasons that caries are still prevalent may be because of the increase in sugar consumption and the continuation of alcohol and tobacco use (especially in the developing world); these are risk factors for tooth decay, periodontal disease, and tooth loss.

Caries and poverty are correlated (Tickle et al. 2000). One in four poor children in the United States has untreated caries. Poverty can lead to caries risk as a result of protein deficiency, which can lead to growth deficiencies (Psoter et al. 2005). Poor nutrition may lead to enamel hypoplasia, which are grooves on teeth that occur when there is a decrease in the quality of enamel being produced, and enamel hypoplasic teeth are more prone to caries than teeth with healthy enamel (Hong et al. 2009; Psoter et al. 2005). Furthermore, poor nutrition may reduce the formation of secondary and tertiary dentin. Maternal vitamin D deficiency can also lead to enamel hypoplasia in infants; mothers who live in urban environments or are prevented from going out or exposing skin due to religion may be especially at risk of vitamin D deficiency. Other risk factors for caries that may intersect with poverty include lead exposure and secondhand smoke exposure (Aligne et al. 2003; Moss et al. 1999); both of these toxins are more frequent in poverty-stricken environments, especially in the United States. Lead exposure has been linked to an increase in caries in a large sample of twenty-five thousand from 1988 to 1994 (Moss et al. 1999). Nutrition also usually varies with socioeconomic status; it appears that poorer families consume high-calorie, low-nutrition foods. For example, Burt et al. (2006) found in a sample of over two thousand that among African-Americans who were well below the poverty level, soft drinks were a major cause of caries. Johansson et al. (2010) found that caries were linked to children's snacking habits; with a sample of over 1,200, they discovered that snacks, such as chips and sweets, increased caries, whereas fresh fruit, crackers, and yogurt did not increase caries.

Recently, the trend of decreasing caries has started to reverse in children according to data from the CDC's NHANES data set (Wang et al. 2012b). Han et al. (2010) found that Koreans had caries rates of over 80%, which they linked to sugar consumption, since sixteenth- to eighteenth-century AD Koreans had caries rates as low as 4%. Another factor that may relate to the increase in caries is the increase in bottle-feeding infants. Carious lesions that develop quickly and involve numerous teeth, and teeth that are not usually affected by caries can be diagnostic of baby bottle tooth decay, which has become a common problem worldwide as mothers go outside the home to work (Barnes et al. 1992; Johansson et al. 2010). The increase in bottled water consumption may also be a factor in the increases in caries because bottled water does not contain fluoride (Horowitz 2003; Lalumandier and Ayers 2000).

Although caries may be largely preventable, there are risk factors unrelated to diet, oral hygiene, and dental visits. Twin and familial studies suggest that caries heritability ranges from 35% to 70% (Wang et al. 2012b). Furthermore, genetic studies on large Iowan samples found that multiple single nucleotide polymorphisms (SNPs) can decrease or increase the risk of caries (Shaffer et al. 2012; Wendell et al. 2010). The genetic components to caries risk include saliva composition and flow, tooth morphology, taste preferences, and enamel formation (Wang et al. 2012b).

Another risk factor involved in caries is dry mouth. Dry mouth has increased in modern Western populations recently as a result of medicines (Parker-Pope 2000; Thomson 2005). Over five hundred drugs list dry mouth as a side effect; medicines for depression, blood pressure, smoking cessation, diabetes, and acne all can lead to dry mouth (Parker-Pope 2000).

Caries may seem like a minor health ailment, but it is the main cause of tooth loss (Efe et al. 2007; Palubeckaitė-Miliauskienė and Jankauskas 2007; Zinoviev 2010). As recently as 1939, tooth loss was a problem even in young adults; for example, the most common cause of World War II draft rejections was too few teeth because of tooth decay (NIH 2013).

PERIODONTAL DISEASE

Periodontal disease is a group of inflammatory conditions that affect the gums and bones that hold teeth in the jaws (Vodanović et al. 2012). **Alveolar** bone loss is the main consequence of periodontal disease, which in time results in tooth loss (Pētersone-Gordina and Gerhards 2011). Microbial organisms cause periodontal disease; these organisms derive nutrients from their human hosts (Petersen and Ogawa 2012). In essence, periodontal disease, which can be called **gingivitis** if it affects only the soft tissue or **periodontitis** if it affects the bone, occurs as a result of bacteria that accumulate and create a biofilm that is referred to as plaque and after hardening as **calculus** or tartar (Petersen and Ogawa 2012; Vodanović et al. 2012). In the 1960s it was thought that calculus caused periodontal disease, but now dental scientists understand that it is bacteria that actually form the plaque and therefore cause periodontal disease (Whittaker et al. 1998). Gingivitis does not necessarily lead to periodontal disease and tooth loss. Periodontal disease has affected humans throughout time, and there is evidence of periodontal disease in a *Homo heidelbergensis* skull, which dates back to over half a million years ago. Several factors may increase the risk of periodontal disease, such as osteoporosis, pregnancy, type II diabetes, immune diseases, and malnutrition. Periodontal disease also seems to be related to systemic diseases, such as cardiovascular diseases.

Determining the presence of periodontal disease in the clinical setting usually consists of looking for calculus buildup, looking for periodontal pockets, or probing the gums to see if bleeding occurs (Petersen and Ogawa 2012). Periodontal disease in skeletal remains is noted by calculus accumulation left on the teeth, porosity or pitting near the tooth socket, bone resorption at the tooth socket, and abscesses (Vodanović et al. 2012). Abscesses are sites where **pus** drainage occurs. When an infection occurs, white blood cells travel to the site to attack the infectious agent, which creates pus, and the pus needs to come out to remove the infection. The hole where the pus exudes is the abscess. Figure 7.4 displays a molar abscess. Taking these traits into consid-

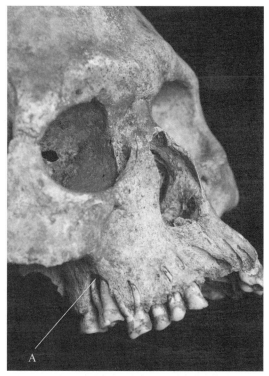

Figure 7.4. Abscess. This twenty-one- to thirty-year-old male has a molar abscess (A), which could have led to tooth loss. Photograph by Daniel Salcedo.

eration, anthropologists have found that periodontal disease can be found in many skeletal samples, such as sixth- to tenth-century Croatia (Vodanović et al. 2012), northwest Canadians from over one thousand years ago (Skinner et al. 1988), Latvian samples of the seventeenth and eighteenth centuries (Pētersone-Gordina and Gerhards 2011), and pre-Columbian Chilean agriculturalists (Meller et al. 2009). There does seem to be a link between agriculture and periodontal disease; Meller et al. (2009) and Cucina and Tiesler (2003) found that oral health declined with the introduction of agriculture in South America. Cucina and Tiesler (2003) suggested that extensive calculus formation as a result of less abrasion in post-contact Peru led to an increase in gingivitis and periodontitis. Meller et al. (2009), looking at a Chilean Atacama desert population, attributed the high rates of caries and periodontal disease to maize dependency.

In modern populations, positive correlations between aging and calculus and gingivitis have been noted (Whittaker et al. 1998). Plus, gingivitis is common in modern populations, especially the elderly, but it is less likely to progress to periodontal disease when compared to past populations (Pētersone-Gordina and Gerhards 2011; Whittaker et al. 1998). Perhaps the lower severity is associated with an increase in nutrition and dental care (Pētersone-Gordina and Gerhards 2011). Sometimes the relationship between calculus and periodontal bone loss, which would suggest a progression from gingivitis to periodontitis, is difficult to find even in non-modern samples; for example, in AD 400, Romano-Britons had higher calculus rates than an eighteenth-century British population, but there was no difference in bone loss (Whittaker et al. 1998). Although past populations may have had a lack of oral hygiene and oral dental care, modern populations have many new risk factors, and some risk factors continue to plague human health.

According to the CDC (2013), nearly half of all thirty-year-old and older individuals in the United States have periodontal disease. The rate of periodontal disease increases with age; by age sixty-five, 70% of individuals have periodontal disease. There are many new risk factors for periodontal disease, such as type II diabetes, smoking, and oral contraceptives. Immune systems are important in the elimination of inflammation found in periodontal disease. Furthermore, some individuals are more prone to infection than others. Crooked or crowded teeth may prevent effective oral hygiene and thus allow bacteria to produce tartar buildup on teeth.

Smoking is perhaps the best-documented risk factor for periodontal disease. Ojima et al. (2007), using a sample of 1,314 Japanese individuals, found that even young smokers between twenty and thirty-nine years of age have more tooth loss related to caries and periodontal disease; however, they found that caries led to more tooth extractions than periodontal disease did. Hanioka et al. (2011) examined literature on smoking and periodontal disease and reported that there is significant evidence that smoking actually does cause periodontal disease. Furthermore, they propose that tobacco suppresses the immune system's functionality, and this leads to an inability to fight off infections in the mouth. Jette et al. (1993) found that duration and amount of tobacco use were directly linked to a decrease in oral health that included tooth loss, tooth decay, and periodontal disease. Smoking may account for half of the cases of periodontal disease in American adults (Petersen and Ogawa 2012). Smoking is even more common in many developing countries. **Betel** chewing, which is a traditional tobacco product in Asia and Oceania, can also cause periodontal disease (Corbet and Leung 2011). Although osteoporosis has been correlated with periodontal disease, research by Megson et al. (2010) suggested that the relationship is confounded with aging and smoking.

Diet can also affect periodontal disease; scurvy, for instance, which is a result of a lack of vitamin C, has been correlated with periodontal disease (Pētersone-Gordina and Gerhards 2011; Petersen and Ogawa 2012). Diets high in sugar have also been found to increase periodontal risk as well as caries risk (Cinar et al. 2013). And in a sample of over thirty thousand males, diets low in fruits and vegetables have been associated with diabetics who have periodontal disease (Jimenez et al. 2012). Type II diabetes and obesity have been correlated with periodontal disease, but it is difficult to determine whether the driving force between the relationship is diet or diabetes.

Other lifestyle choices can also affect periodontal disease. When agriculture was adopted, female risk of periodontal disease increased, which may have been linked to increases in fertility (Watson et al. 2010). Settling down and having extra calories available allowed females to have offspring more frequently. During pregnancy, hormonal changes alter the immune response system, and the body's inflammatory response is lowered (Wandera et al. 2009). This decrease in immuno-responsiveness increases the susceptibility to bacterial infections, and what may have stopped at gingivitis is more likely to progress to periodontitis. Periodontal disease increases the risk of premature delivery, and thus women should be especially attentive to their oral hygiene during pregnancy (Clothier et al. 2007; Wandera et al. 2009). Even without pregnancy, females may be putting themselves at risk of periodontal disease. Oral contraceptives, which alter hormones in ways that mimic pregnancy, may also put one at risk of periodontal disease.

Although periodontal disease is extremely pervasive in the modern world, toothbrushing is the most effective treatment for the disease (Cinar et al. 2013). Even individuals who engage in risky behaviors, such as smoking, can reduce their risk with toothbrushing. The main reason to treat periodontal disease is to prevent tooth loss.

TOOTH LOSS

Tooth loss, which leads to bone resorption at the jawbone, has a multifactorial etiology, but caries and periodontal disease are the most common causes for tooth loss today (Chen and Clark 2011). In non-human primates, tooth loss is rare (Gilmore 2013), but tooth loss occurred early in human evolution. The remains from a 1.77-million-year-old *Homo erectus* from Dmanisi, Georgia, present some of the earliest evidence of tooth loss with bone resorption, which proves that the tooth loss occurred prior to death (Russell et al. 2013). Another early case of tooth loss is from the Old Man Neanderthal (who was likely only forty years of age) found in France, which dates to about sixty thousand years ago (Russell et al. 2013). Anthropologists call tooth loss prior to death antemortem tooth loss, and the evidence is when the tooth socket is closed or closing as a result of bone resorption. Figures 7.5 and 7.6 show tooth sockets that have closed up through bone resorption. In clinical studies, tooth loss is easy to assess; dentists can obviously see when teeth are missing, but even individuals can report the data on their own tooth loss or retention.

Bioarchaeologists have found that tooth wear and trauma were also likely risk factors for tooth loss among hunter-gatherers (Russell et al. 2013). It appears that the introduction of agriculture led to an increase of caries and tooth loss related to caries (Larsen 1995; Russell et al. 2013), but this trend is

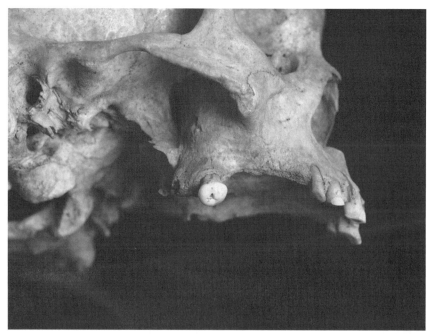

Figure 7.5. Antemortem tooth loss. This female who was between thirty-nine and forty-four years of age lost nearly all of her teeth prior to death. Her third molar is still in place. Photograph by Daniel Salcedo.

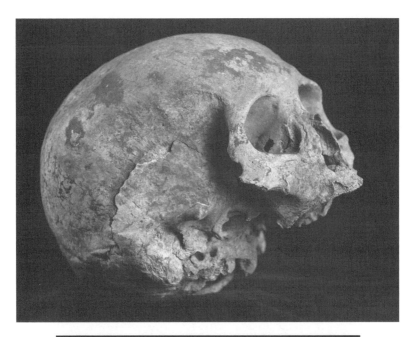

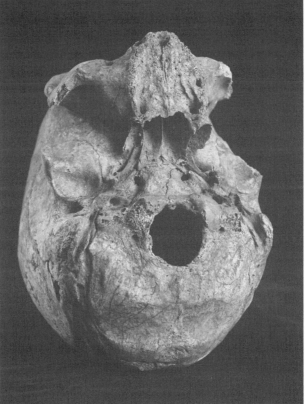

Figure 7.6. Edentulous. A lateral view (top) and inferior view (bottom) of a female hunter-gatherer who was over thirty-five years old and had lost all her teeth, which is also called edentulous. Photographs by Daniel Salcedo.

not found in the Old World (e.g., Ubelaker and Pap 2009). And even in some New World sites, agriculture does not lead to an increase in tooth loss (e.g., Watson 2008b). Use of teeth as tools and ritual tooth removal have also been documented by anthropologists (Russell et al. 2013).

More recently, tooth loss was considered a normal part of aging. As recently as the 1950s, tooth loss was extremely common (Müller et al. 2007). Barnes et al. (2002) examined remains from the Hamann-Todd Collection compared to living patients at free clinics and provided evidence that even the poorest populations in the present-day United States receive better dental care and lose fewer teeth today than they did seventy years ago. In short, the free clinic patients, Barnes et al. reported, had more teeth and more fillings than the Hamann-Todd Collection individuals. In the Hamann-Todd Collection, the great majority of older people were toothless (or **edentulous**). Edentulous rates have declined rapidly in just the past few decades; for example, in 1972 in Scotland 44% of all adults were edentulous, but by 1998 the rate dropped to 18% (Müller et al. 2007). In England and Wales in 1968, 22% of adults between the ages of thirty-five and forty-four years old were edentulous; thirty years later nearly 100% of forty-five-year-olds had some teeth, but by age fifty-four it is still common to be toothless (Steele et al. 2000). Nonetheless, there is a great deal of variation in dentition retention; for example, Müller et al.'s meta-analysis of Europe revealed edentulous rates between 3% and 80% among individuals over sixty years of age. Even in countries that are similar to one another, such as Sweden, Denmark, and Finland, variation can be great; Sweden's edentulous percentage of people over seventy-five years old is 27%, while the rate is twice as high in Finland (Müller et al. 2007). In general one may assume that females retain fewer teeth than males since they are more prone to periodontal disease as a result of hormones, but the literature shows conflicting results (Müller et al. 2007; Russell et al. 2013).

Several demographic factors shape who retains their teeth and who does not within countries. For example, regardless of location, individuals with lower socioeconomic statuses have greater tooth loss (Avlund et al. 2011; Gilbert et al. 1993; Müller et al. 2007). In the United States, tooth loss in whites is associated with lower incomes, but black and Hispanic populations are not as greatly affected by their income (Jimenez et al. 2009; Russell et al. 2013). Yet, overall, whites have greater tooth retention than blacks and Hispanics (Russell et al. 2013). The reasons for the link between socioeconomic status and tooth loss are multifactorial. Poorer people may not seek dental care as often (e.g., Gilbert et al. 1993). Nutrition may also be a factor; vitamin C deficiency, for example, is related to tooth loss (Lowe et al. 2003), and dairy consumption is related to tooth retention (Adegboye et al. 2012). Poor nutrition in the developing nations has been linked to greater tooth decay, which may in turn cause greater tooth loss (e.g., Efe et al. 2007). Even in developed nations, poverty-stricken individuals often eat high-sugar, low-nutrient foods. Lead, which is found in older pipes, in candy from developing countries, and in old paint, has been linked to bone remodeling disruption and salivary changes that can cause tooth loss. In a study of 2,280 US veterans, Arora et al. (2009) found that lead exposure increased tooth loss as much as cigarette smoking does. Lead exposure is greatest in poor regions and in developing countries.

The vast majority of tooth loss is related to poor dental hygiene and smoking, which are both risk factors for periodontal disease and caries. For example, Jansson et al. (2002) followed individuals for twenty years to determine the greatest predictors of tooth loss; in their sample of 513 individuals from Stockholm, they noted that plaque, periodontal disease, and smoking led to lost teeth, but smoking without periodontal disease did not correlate to tooth loss. Smoking is a risk factor for periodontal disease. Gilbert et al. (1993) examined senior Floridians and found that tooth loss was mostly a result of poor dental hygiene and smoking; both of these factors can lead to caries and periodontal disease. Oral hygiene, especially flossing, has been tied to a reduction in periodontal disease; Huang et al. (2013) found a relationship between the percentage of teeth present and flossing.

Although the causal relationships are not fully understood, obesity and type II diabetes correlate with periodontal disease and tooth loss. Östberg et al. (2009) examined a Swedish population and found that higher BMIs and greater abdominal circumferences were associated with greater tooth loss, but only in individuals under sixty years of age. In Britain, the correlation remains in elderly individuals. The likelihood is that the higher BMIs and abdominal circumferences relate to a diet high in sugar.

Hayasaka et al. (2013) estimated that when one has ten teeth or less, the individual is at a three times greater risk of health problems. And Avlund et al. (2011) found that in a sample of 734 Danes, tooth loss was related to fatigue in individuals seventy years old or older. However, in many of these studies, smoking is confounded with disease, such as strokes and cardiovascular disease (e.g., Lowe et al. 2003; Peres et al. 2012). Clinicians theorize that loss of teeth may relate to poor health because edentulous patients eat less nutritious diets. For example, in a study of over twenty thousand Japanese dentists between the ages of twenty-six and ninety-eight years old, Wakai et al. (2010) found that tooth loss led to a decrease in vegetable consumption. Additionally, clinical researchers have suggested that infections may cause inflammation of the heart muscles and thereby cause cardiovascular disease (Kaye et al. 2010; Polzer et al. 2012). Regardless of cause-and-effect relationships, good oral hygiene (that includes flossing), good nutrition, and abstaining from smoking can greatly improve the odds of retaining your own teeth.

MALOCCLUSION

Malocclusion is a term used to refer to the misalignment of dentition. Misalignment can be a result of dental over-crowding or an abnormal relationship between the maxilla and the mandible. Malocclusion can be seen in an examination of a patient or through the use of X-rays. In anthropological studies, malocclusion requires well-preserved remains that have both upper and lower jaws intact with the teeth in place. Dentists score malocclusion in three classes. Class one is the most common, and it is when the bite is normal, but the upper teeth overlap slightly with the lower teeth. Class two, which is called an overbite, is when the maxilla and its teeth severely overlap the mandibular jaw and teeth. Class three is an underbite and occurs when the mandible protrudes forward and thereby the mandibular teeth overlap the maxillary teeth and jaw. Figure 7.7 shows an overbite.

Although malocclusion can result from tooth wear and tooth loss, most malocclusion discussions in anthropology have been in regard to malocclusion from a mismatch between tooth and jaw size. Normando et al. (2011) commented that the etiology of malocclusion is unclear, but anthropologists have defined it as a "disease of civilization" and blamed processed foods for the increase in malocclusion. One must first ask, however, whether an increase in malocclusion has occurred through time.

Malocclusion rates in modern populations range from 40% to 80% (Daragiu and Ghergic 2012; Evensen and Øgaard 2007); before the nineteenth and twentieth centuries, it appears that malocclusion was rare. Yet there is a Neanderthal with malocclusion dated to one hundred thousand years ago, and missing molars have been reported in skulls dating over ten thousand years old. Evensen and Øgaard (2007) found a low rate of severe malocclusion in a sample of medieval skulls from Oslo compared to modern Norwegians. In their research, they found that 65% of modern Norwegians with malocclusions needed treatment, whereas medieval skulls would have required treatment only 36% of the time. Lavelle (1976), looking at a large sample of modern British people, medieval British people, Anglo-Saxons, and modern Africans and Asians, found that the modern British population had the most over-crowding. In Japan, malocclusion rates have been traced from 1000–500 BC (i.e., hunter-gatherer Jomon) to modern Japanese (1964–1966). The Japanese Jomon hunter-gatherers had malocclusion rates of about 20%, whereas agricultural Japanese had rates of malocclusion that were close to 50%; the modern Japanese malocclusion rate is 76% (Larsen 1997). This increase has been tied to a softer diet that prevents strong jaw growth. Other cases of increased malocclusion with a softer diet come from Pima Indians (Corruccini et al. 1983). Larsen (1995) reviewed the literature on malocclusion in relation to the adoption of agriculture and found that dietary changes led to an increase in malocclusion throughout the Americas. In short, those study-

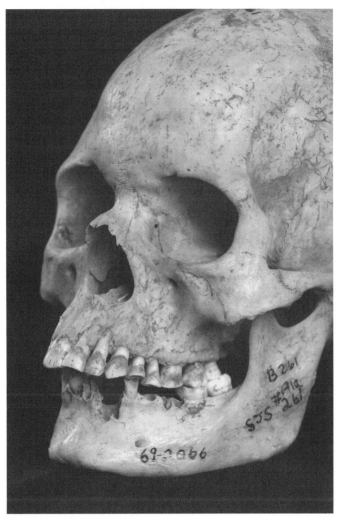

Figure 7.7. Malocclusion. Although usually associated with modernity, even in prehistoric samples malocclusion occurs, and perfect occlusion is rare. Here is an example of a pre-contact California Amerind with an overbite. Photograph by Daniel Salcedo.

ing rates of malocclusion in the past compared to agricultural or industrial societies have suggested that the ease of processed food consumption reduced jawbone size (by reducing the stresses on the bone and thereby reducing bone remodeling), but tooth size is more firmly controlled by genetics and thus did not change much. Some studies on the Hamann-Todd Collection, the Terry Collection, and the Forensic Anthropology Database have found that mandibles have actually been getting longer from the early twentieth century to the later twentieth century (Martin and Danforth 2009). Martin and Danforth suggested that the change is a result of better nutrition. Not all researchers are convinced by the environmental argument for malocclusion (see McKeever 2012). Normando et al. (2011) found malocclusion in Amazon tribes and linked it to genetics. The two tribes examined included a large diverse village and a village that resulted from a group of individuals who splintered off from the larger population. The smaller village had an increase in inbreeding as a result of its size. Malocclusion rates were 33% in the large village and 63% in the smaller village, which the authors attribute to inbreeding.

Determining the rate differences and causes of malocclusion can be difficult because different measures are taken, dry bone results may differ from living individuals, and tooth loss and wear can affect occlusion and the ability to diagnose malocclusion. Regardless of the cause, malocclusion results in negative health effects. Hörup et al. (1987) used a sample of nearly five hundred Danes and discovered that individuals who are unhappy with their bites are more likely to have lower dental hygiene. Malocclusion was also found to decrease oral hygiene in Bucharest (Daragiu and Ghergic 2012).

Malocclusion may lead to third molar impaction. Third molars, which are also known as wisdom teeth, are the most commonly extracted teeth. Third molar extraction is the most common type of oral surgery (Martin et al. 2005). Often surgery results in inflammation that can last for several months (Cei et al. 2012); surgery can also cause infections, **necrosis** (death) of bone and tissue, and nerve and tissue damage that may be permanent (Friedman 2007; Martin et al. 2005). Even though complications occur, the dental community has continued removing third molars even when they are not symptomatic (Friedman 2007). Third molars are extracted for a variety of reasons. In the early years of modern dentistry before **antibiotics**, molars were removed because they were often thought to be associated with infections that could lead to death (Pratt et al. 1998). Antibiotics made minor infections treatable without molar extraction. Starting in the 1980s, research on molar extraction has found that between a quarter and a half of all third molars extracted were not extracted for valid reasons (Pratt et al. 1998). Frequent reasons for molar extractions include elective extraction, caries, pain, and impaction (Ong et al. 1996; Martin et al. 2005). Friedman (2007) has questioned whether impaction would actually occur in some of these third molars. Early treatment has resulted in trying to predict impaction prior to tooth eruption (Friedman 2007). Dentists have suggested that pressure from erupting third molars can result in dental crowding, especially of the anterior teeth. Yet Friedman argued that molars erupt in spongy cancellous bone and cannot push the firmly rooted anterior teeth. Furthermore, he suggested that half of the third molars classified as impacted will actually develop normally.

CONCLUSIONS

In many ways, dental health decreased with the onset of agriculture and industrialization; lower levels of occlusal wear meant higher caries, more calculus, and fewer teeth. However, recently, just in the last couple of decades, dental health has improved greatly. Fluoridation of water and sealants reduced caries, use of dental products such as toothbrushes and floss have reduced periodontal disease, and accessible fruits and vegetables all year round likely led to reduced tooth loss. Other modern conveniences and luxuries, however, do reduce dental health. For example, the excessive amount of sugar available can be seen as one reason for the reverse trend in caries reduction. Substituting sugared drinks for milk and snacking between meals probably also add to the reverse trend in caries reduction. Finally, the improvement in dental health is uneven; developing nations are still lacking fluoride in their water and sometimes even lacking fluoridation in their toothpaste. Plus, as fewer people in the developed nations smoke as a result of the war on cigarettes, the tobacco companies will likely continue their profits in developing nations.

CHAPTER 8

Infectious Diseases

According to the NIH (2013), "infectious diseases kill more people worldwide than any other single cause." Infections come in four types: (1) bacterial, (2) viral, (3) fungal, and (4) protozoan. **Bacteria** are one-celled organisms that multiply quickly and release chemicals that may make the infected individual ill. **Viruses** are capsules that contain genetic information and require host cells to multiply. **Fungi** are primitive plants, and a **protozoan** is a one-celled animal that uses other living organisms for food and a place to reproduce and thrive. An infectious disease is considered contagious if it is spread from one person to another person.

Bones are usually protected by soft tissue, but infectious agents can enter the system either through a fracture, through the circulatory system, or through the lymphatic system. A common entry for pathogens is through skin **ulcers**, which are more common in the tropics, in malnourished individuals, and where poor hygiene exists (Boel and Ortner 2013). In diabetics, the feet are particularly prone to ulcers. However, tibiae are the most likely bone to be initially infected due to the thin skin barrier to the bone coupled with the large medullary cavity and large foramina on the tibia. Bone-forming lesions of the tibia are one of the most common abnormalities in human skeletal remains (Boel and Ortner 2013). Bacteria and viruses that are inhaled (such as pneumonia or tuberculosis) or consumed (such as in *Salmonella* and *E. coli*) are also common in current populations and were likely common in prehistoric populations. Parasitic worms, which are called **helminths**, that can be ingested or that enter from the skin are less common in developed countries, but they still affect millions of individuals in developing countries.

Most infectious diseases do not leave any trace of their existence on bones, and even fewer infectious diseases leave distinct marks on bone. For example, *Salmonella*, which can cause typhoid fever, is common in developing countries, but bony evidence of infection from *Salmonella* occurs in less than 1% of afflicted individuals (Mathuram et al. 2013). Skeletal effects from tuberculosis occur in only about 3% to 6% of infected individuals (Gadgil et al. 2012). Infectious agents may cause quick death in the weakest individuals, and thus the bone is not affected. And chronic infections that do cause bony changes may actually be a sign of good health since a weak individual may die faster than a healthier individual.

Anthropologists who study infections on skeletal remains may be able to identify some specific infectious diseases, such as tuberculosis, syphilis, and leprosy, but for the most part, determining the cause of the infection is unsuccessful. Many of the diagnostics rely on which bones are afflicted, but assumptions that these infectious diseases only affect specific bones are often erroneous. Thus, the possibility for diagnostics in the archaeological record really relies on ancient DNA; for example, Mutolo et al. (2012), examining tenth- to thirteenth-century Albanians, found that vertebral changes that could have been attributed to tuberculosis, *Brucellosis*, or hernias were actually determined to be *Brucellosis* (which is a bacteria transmitted from animal products, especially unpasteurized dairy products, that causes fever, muscle aches, fatigue, headache, and night sweats, but rarely causes death) through DNA testing.

Even with all these complications, the frequency and trends of infections can aid in reconstructing past populations' health, especially with regard to temporal trends. Mitchell (2003) reviewed the changes in pathogen forms and rates over time and space. He found that although infectious diseases have been part of the human experience since the beginning of **hominins** (around six million years ago), domestication of wolves fifteen thousand years ago led to the transfer of various diseases from canines to humans. Human-induced changes in landscape can also result in an increase or decrease in pathogen infections. In Egypt, for example, the irrigation of the Nile valley led to an increase in helminth infections (such as those from ***Schistosoma***, a parasitic worm that can cause damage to the liver, bladder, intestines, and lungs) as far back as 300 BC; infections again increased with the construction of the Aswan Dam (Hibbs et al. 2011). Agriculture, livestock, and irrigation ten thousand years ago led to widespread endemic infections. Cities at around six thousand years ago led to epidemics, and not until two hundred years ago did **pandemics** arise. The rise in infectious disease outbreaks seems to be correlated with population density and unsanitary conditions. Bacteria, viruses, fungi, and helminths can travel easily from one individual to the next when there are more people living close to one another. Furthermore, if sanitation is lacking, then infectious agents can travel through the water or sewer system. Many bacteria are expelled through feces; if feces get into the water system, then it may travel and infect individuals who utilize the water source. Scarce resources, poor diet, and stress can affect an individual's immune system and their ability to fight off the infection.

GENERAL BONE INFECTIONS

Bone infections may arise from soft tissue infections. Bacteria, viruses, parasites, and fungal infections spread easily through the bloodstream and can result in irreversible damage that kills bone cells, which is called necrosis. Patients with bone infections (**osteomyelitis**) may experience redness, warmth, stiffness of joints, and pus around the infected area. Systemic symptoms also include chills and fever. Doctors may also take into account risk factors, such as diabetes, rheumatoid arthritis, **hemophilia, sickle-cell anemia**, and HIV. Clinicians will also use X-rays and MRIs to search for evidence of infection. Osteomyelitis can be caused by a variety of infectious agents, such as *Fusiform bacilli*, **Streptococci** *sp.*, and *Mycobacteria*, but the most common cause of osteomyelitis is *Staphylococcus aureus* (Boel and Ortner 2013; Santos and Suby 2012). Lab tests of bone tissue and blood samples may also reveal the **pathogens** that are causing the infection, but blood tests can show a lack of infection in 40% of cases even when infections are present (Wedman and van Weissenbruch 2005).

Bone Infection Diagnostics in Skeletal Remains

Looking at patterns of bone involvement may help bioarchaeologists determine which infectious agent may be responsible for the osteomyelitis. For example, tuberculosis is usually thought to affect the vertebrae and ribs; however, tuberculosis can also be found in the knees, mandibles, and other bones (Agadi et al. 2010; Gadgil et al. 2012). To further complicate matters, some infectious diseases may mimic trauma; for instance, Mays (2007) suggested that many cases of *Brucellosis* are diagnosed through anterior vertebral body erosion, but the vertebral body erosion is similar in appearance to hernias. **Smallpox**, which was eradicated by the WHO (2013), with the last official case being reported in 1977, causes fusion and erosion of the elbow, which can mimic traumatic arthritis (Darton et al. 2013). Furthermore, infections often do not affect bones and thus may be missed in the bioarchaeological record. Osteomyelitis from smallpox occurred rarely, and there are few cases reported in the bioarchaeological record, although many people were inflicted with the disease. Two known cases have been identified: a tenth- to twelfth-century French teenage male (Darton et al. 2013) and an AD 1640 to 1650 Native Canadian adult male (Jackes 1983). Although looking at bony variations to find out if individuals experienced an infection is the most common form of diagnostics in bioarchaeology, histology and DNA analyses are becoming more common. Poliomyelitis (abbreviated as polio), a viral disease that is contagious and incurable, causes paralysis in about one in two hundred infected individuals (Gholipour 2013). In historic times, polio rose to the level of an epidemic; two cases of polio from historic times include a medieval Austrian female and a nineteenth-century African-American male from Mississippi (Thompson 2012). Earlier cases of polio have been hard to document due to the unreliable skeletal diagnostic traits, and even clinical data suggest variability of symptoms in polio, but in ancient Egypt, both artistic depictions and skeletal remains are suggestive of polio's presence (Thompson 2012). **Histological analyses** cannot necessarily help in diagnosing skeletal infectious diseases (van der Merwe et al. 2010), and DNA analyses may be ineffective if contamination has not been controlled for.

Bioarchaeologists use the term **osteitis** for a general bone infection; they also document two different types of bone infections in skeletal remains: periostitis and osteomyelitis. Figures 8.1 and 8.2 display osteitis on the lower limb. Periostitis, which is also known as a periosteal reaction, is usually localized to one or two bones and is rarely fatal. Periostitis occurs when an insult causes the elevation of the outer layer of the periosteum, which is due to compression and the stretching of blood vessels as a result of an infectious agent. Weston (2008) questioned the legitimacy of regarding all periosteal reactions as being indicators of infection. She examined periosteal reactions of museum specimens and found that reactions were often the result of trauma and bone remodeling without inflammation or infection. Thus Weston argued against defining periosteal reaction as an indicator of infection; the medical community seems to ignore periosteal reactions and focuses on osteomyelitis instead. It appeared to Weston (2008) that periostitis should only be considered infectious when other features of infections, such as cloacae, are present. **Cloacae**, which are pus-draining holes, are a common symptom of severe bone infections; a cloaca is seen on the humerus pictured in figure 8.3. **Pyogenic** agents, which are bacteria, viruses, fungi, or helminths that cause pus, will result in cloacae. Pus, as mentioned in chapter 7, is the result of white blood cells fighting off the infection and needs to exit the site of infection. Severe infections, such as osteomyelitis, nearly always involve cloacae. With osteomyelitis, many bones are usually involved, and the infection spreads through the medullary cavity. Anthropologists note that osteomyelitis likely would be fatal in prehistory. According to a study of over five hundred remains from Portugal, even in the first half of the twentieth century, which was a pre-antibiotics and pre–modern surgery period, osteomyelitis accounted for 77% of deaths of individuals between eight and fourteen years old (Santos and Suby 2012). Djurić-Srejić and Roberts (2001) examined 1,617 skeletons from eight medieval Serbian cemeteries and found that infection rates were high, and nearly a third of infections led to death in Serbia prior to the use of antibiotics.

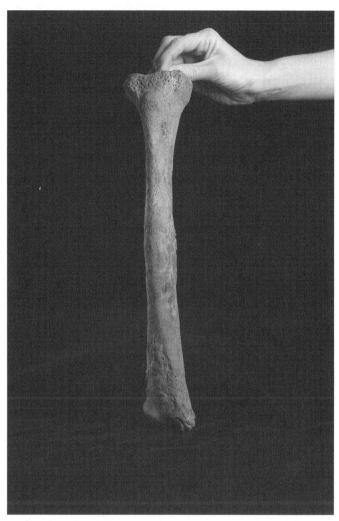

Figure 8.1. Osteitis of a tibia. The roughened exterior of the shaft coupled with the abnormal increase in diameter of the shaft (seen most clearly in the bottom half of the bone) is indicative of bone infection on this tibia. Photograph by Daniel Salcedo.

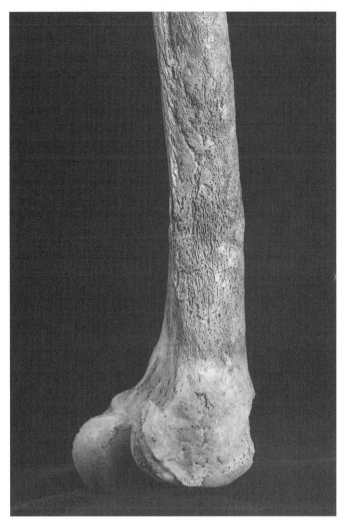

Figure 8.2. Osteitis of a femur. The roughened exterior of the shaft is indicative of bone infection on this femur. Photograph by Daniel Salcedo.

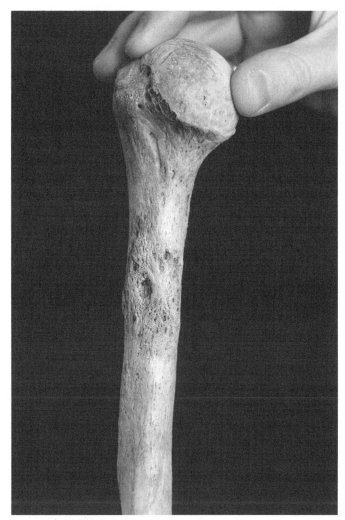

Figure 8.3. Cloaca on a humerus. This thirty-nine- to forty-four-year-old had osteomyelitis; multiple bones had evidence of infection, and in this particular bone one can see the pus-draining hole called the cloaca. Photograph by Daniel Salcedo.

Bone Infection Risk Factors

Anthropologists have taken notice of the increase in osteomyelitis with the advent of agriculture, sedentary behavior, and population growth. For example, Oxenham et al. (2005) found that increased infection rates of North Vietnamese populations from the Bronze and Iron Ages (3,300 to 1,700 years ago) compared to the earlier coast foragers (6,000 to 5,500 years ago) were attributable to the onset of chiefdoms coupled with agriculture and an increase in population density. Similar findings have been documented for the Levant at the transition of hunter-gatherer to agriculture (Eshed et al. 2010).

Clinicians find that children's long bones are the bones most often afflicted with osteomyelitis. The pathogens can travel easily through the bloodstream and enter into the blood-rich growth plates in children. Pathogens from urinary tract infections and pneumonia are common in children. Infections can also result from a nearby injury.

Currently, according to the WHO (2013), 347 million people worldwide have diabetes; 80% of diabetics live in the low- and middle-income countries. Out of the 347 million with diabetes, 90% have type II diabetes. Type II diabetes is the result of the body's ineffective use of insulin, which is a hormone released through the pancreas that regulates carbohydrates and fat metabolism; type II diabetes is avoidable through maintaining a healthy body weight and diet. From the 1970s to the 1990s, type II diabetes rates have doubled, and the increase did not slow in the 1990s and 2000s (Fox et al. 2006). A common diabetic symptom is **neuropathy**, which occurs when blood flow is reduced and nerves are damaged, resulting in foot ulcers that become easily infected. Over 15% of the sixteen million type II diabetic patients in the United States develop foot ulcers and osteomyelitis (Nyazee et al. 2012). Usually the patient's infection can be treated with a one- to one-and-a-half-month antibiotic regimen coupled with **surgical debridement** (Nyazee et al. 2012). And when the foot does not heal properly, infections are not cured with antibiotics, or the patient has waited too long, amputation may become unavoidable (Nyazee et al. 2012). Unfortunately, this treatment is often followed by slowed healing, and although theoretically osteomyelitis should increase bone turnover and remodeling, in practice it does not (Nyazee et al. 2012). Patients often have post-surgical complications; in a small study of ninety-four individuals, Aragón-Sánchez et al. (2012) found that 40% of patients developed new ulcers, which may be a result of new pressure points created by partial amputations that alter the biomechanics of the foot. The most common type of infection in diabetics is a *Staphylococcus aureus* infection, which is bacterial. It can be antibiotic resistant and then is referred to as methicillin-resistant *Staphylococcus aureus*, or MRSA for short. Some studies have found that about a third of infected diabetic patients had MRSA (Aragón-Sánchez et al. 2009).

Other surgeries can also result in osteomyelitis (Rod-Fleury et al. 2011). Wang et al. (2007) found *Mycobacteria tuberculosis* in a sternum of a fifty-seven-year-old Chinese female who had undergone coronary artery bypass grafting. Spinal infections can occur from spinal surgeries; for instance, a twenty-nine-year-old patient who was operated on for a pinched nerve developed a *Candida albicans* infection from an accidently perforated bowel (Quesnele et al. 2012). And infections occur in between 0.3% and 1.7% of anterior cruciate ligament reconstructions, which are common elective surgeries (O'Neill et al. 2013; Sun et al. 2012). Other medical interventions, such as the use of oral bisphosphonates to treat osteoporosis, can also lead to osteomyelitis. Yamazaki et al. (2012), using a sample of nearly eight thousand, found that oral bisphosphonates increased osteomyelitis of the jaw by five to six times compared to other drugs to treat osteoporosis.

PARASITE-INDUCED ANEMIA

Iron deficiency is the most common nutrition problem in the world (Wright and Chew 1998). Iron is needed for transporting oxygen to blood and body tissues. When the body is iron deficient, also known as anemic, normal motor functions and thought processes can be delayed. Anemia decreases the ability to work and can even affect the functioning of vital organs.

In clinical samples, blood tests can indicate iron deficiency; hemoglobin (Hb) concentration of over 110 grams per liter (g/l) indicates no lack of iron; between 90 and 109.9 Hb g/l is mild iron deficiency; 70 to 89.9 Hb g/l is moderate iron deficiency; and below 70 Hb g/l is considered severe iron deficiency (Seck and Jackson 2010). As mentioned in chapter 2, the two forms of iron deficiency skeletal markers are cribra orbitalia and porotic hyperostosis. Cribra orbitalia occurs before porotic hyperostosis; thus porotic hyperostosis is indicative of prolonged childhood anemia (Wright and Chew 1998). Cribra orbitalia and porotic hyperostosis are a thinning of cranial bones that result in a hair-on-end appearance of bone when X-rayed. The thinning bone, which occurs prior to growth completion, is a result of hyperactivity of red blood cells, which causes marrow spaces to fill up with red blood cells rather than other bone cells. The excess red blood cells result in bone expansion without good bone cell formation. When examining a dry skull, such as in an archaeological setting, these features are visible as pinpoint holes, or if the anemia was active at death, a rough and very porous appearance. Since these traits occur during growth and development, they present an indicator of stress during childhood and not adulthood.

Scientists researching iron deficiency have put forth three main causes: (1) anemia is the result of a lack of dietary iron or a lack of other nutrients that enable iron absorption; (2) anemia is actually an evolutionary adaptation to pathogens; or (3) parasites, rather than diet, cause anemia. There are foods that are iron inhibitors, such as maize, and diets that are low in iron, such as those based mainly on unfortified cereals (see chapter 2). Diets low in iron or high in iron inhibitors can result in iron deficiency. Conversely, some researchers have suggested that iron deficiency may actually be an evolutionary adaptation (see Holland and O'Brien 1997 and Wright and Chew 1998). The hypothesis submits that suppressing iron makes it difficult for pathogens to grow and reproduce; ergo, low iron will enhance immunity to pathogens. This adaptive argument suggests that diet only plays a minor role in iron deficiency. Furthermore, animal research has shown that iron is withheld from blood cells when bacteria, such as tuberculosis, have invaded the animal (Wright and Chew 1998). The problem with the adaptive hypothesis is that iron deficiency is extremely deleterious; for example, iron-deficient children are more likely to be hospitalized for gastrointestinal and respiratory infections (Peckmann 2003). Iron deficiency also increases infant mortality and reduces physical and intellectual abilities (Holland and O'Brien 1997).

Many protozoan parasites, such as hookworms, feed on blood and thereby remove iron from the body. Hookworms can remove between 0.05 ml to 0.2 ml of blood per day; and in South Africa, which is considered part of the hookworm belt, between 40% to 100% of people in various villages have hookworms (Holland and O'Brien 1997; Peckmann 2003). Other helminths, such as roundworms, diminish the absorption of iron (Wright and Chew 1998). And whipworms and hookworms also cause intestinal bleeding and bloody diarrhea (Wright and Chew 1998). It seems that parasites can induce iron deficiency and thus result in cribra orbitalia and porotic hyperostosis. Archaeological studies that have highlighted the parasitic factor of cribra orbitalia and porotic hyperostosis include data revealing trends of rising anemia with increased sedentary behavior and increased population sizes. Cribra orbitalia and porotic hyperostosis are not found in the Paleolithic; these markers seem to make their appearance with the onset of agriculture (Peckmann 2003). The cause of iron deficiency in agriculture is likely the result of a low-iron diet coupled with the increase in parasites as a result of increased population density and poor sanitation (Holland and O'Brien 1997; Wright and Chew 1998).

Marine foods, especially when undercooked, have been linked to parasites and anemia in bioarchaeology. For instance, Šlaus (2008), who worked on a large sample of nearly a thousand individuals from Croatia dating from the Late Antiquity period (third to fifth centuries) and the early medieval period (sixth to tenth centuries), found that the early medieval period population consumed marine resources to a greater extent than the earlier Croatians. This, Šlaus argued, may explain some of the increase in cribra orbitalia. Oxenham and Matsumura (2008) found that prehistoric northern Japanese populations had higher frequencies of cribra orbitalia than other prehistoric cold-adapted populations, which the authors attributed to a high parasite load (which they also observed in Southeast Asians today) due to the consumption of under-cooked and raw fish and marine animal foods. Interestingly, Oxenham and Matsumura found that the prehistoric Alaskan populations who seemed to have a greater dietary variety and ate caribou and elk, which are less likely to be parasite rich compared to marine resources, had the lowest cribra orbitalia rates among the prehistoric populations they examined. Further evidence of marine foods being high in parasites arises from the Pacific Coast. Prehistoric foragers from British Columbia to Southern California had high levels of anemia (from 13% to 32%, and on one island 72%), and their diets were rich in iron from marine resources, but the parasite load may have been high due to the consumption of undercooked foods coupled with high population densities that created unsanitary conditions and contaminated waters (Larsen 1997).

In clinical samples, researchers have also investigated the link between helminths and iron deficiency. The WHO estimates that 3.5 billion people worldwide have intestinal parasites (Hussein 2011). In Bangladesh, Iran, Saudi Arabia, India, Turkey, and Zambia, intestinal parasites infect half of the population (Hussein 2011). With immigration to Europe, Australia, and the United States, intestinal parasites are becoming more common; immigrants in Europe have rates of between 17% and 93% depending on the population (Geltman et al. 2001; Rice et al. 2003). Although parasite frequency in itself may be of interest, the correlation with anemia is of particular interest to anthropologists since osteologists have skeletal evidence of anemia but not of parasites. Researchers have examined the relationship between parasites and anemia using blood and stool samples. Rice et al. (2003) noted that in a sample of 133 East Africans living in Australia, half had parasites, but anemia was uncommon. Also, Haidar (2010) looked at data from nearly 2,300 individuals from Ethiopia and found that the most common parasite was the less invasive *A. lumbricoids*, which is a small roundworm that is not likely to cause anemia. Wördemann et al. (2006) examined Cuban schoolchildren and found in the sample of 1,320 that although over half had parasites, only about a fifth had helminths. Conversely, some studies did find correlations between anemia and parasites. For instance, Magalhães and Clements (2011) examined West African children to determine whether iron deficiency protected against infections and thus whether **deworming** might be more beneficial than nutri-ent supplements. They found in their sample of over seven thousand children that anemia correlates with malaria, hookworm, and *Schistosoma*. They also found that iron-deficient diets were the main predictors of anemia. Thus, to improve child health in West Africa, nutrition supplements that include vita-min A (which increases iron absorption) and iron supplements are important, but deworming is also important. Seck and Jackson (2010) also found that although parasite load can cause anemia, diet is the most important factor in predicting anemia. It appears that there is a synergistic effect with diet and parasites (Wright and Chew 1998). Parasites may not reduce iron enough to cause anemia, but if a child already has low iron, then the extra blood lost from parasites may result in severe anemia. Furthermore, low-protein diets increase susceptibility to infectious disease (Schaible and Kaufmann 2007).

TREPONEMAL DISEASES

Treponemal diseases are bacterial infections that are caused by various species and subspecies of the genus *Treponema* that are commonly referred to as venereal syphilis and endemic syphilis. Treponemal bacteria are virulent and cause chronic disease (Buckley and Dias 2002). The endemic forms are pinta, yaws, and bejel (Roberts 2000). Endemic syphilis occurs in specific regions; yaws is found in hot humid areas, whereas bejel is found in temperate and subtropical arid regions, such as the Middle East (Roberts 2000). Endemic syphilis is spread via skin or mucous contact. According to the WHO (2013), yaws is a neglected tropical disease that affects the skin, bone, and cartilage. It is a contagious disease. Skin ulcers (or chancres) usually appear within two to four weeks of exposure; lesions are usually found on the lower extremities or the buttocks. Symptoms include bone pain, lymph node swelling, and fever. Venereal syphilis, which is spread through sexual contact, is found mostly in urban areas worldwide and can also be passed congenitally. Venereal syphilis results in an ulcer and then a skin rash within several months. Syphilis ulcers, which are usually on the penis, vaginal area, or buttocks, may take ten to ninety days to appear (Ferguson and Varnado 2006). Then the ulcer, which is firm, round, and painless, heals within three to six weeks. The healing of the ulcer marks the end of the primary stage; the secondary stage includes a skin rash and red, rough palms and soles (Ferguson and Varnado 2006). The third stage, which is preceded by a latent period, may result in osteomyelitis (Buckley and Dias 2002; Ferguson and Varnado 2006). In addition, syphilis and the related diseases can attack the central nervous system and cause **psychoses**. Congenital syphilis, which is inherited through the mother **in utero**, is only common in venereal syphilis infections and results in abnormalities of the teeth.

Diagnosing Treponemal Diseases

Skeletal symptoms for all the treponemal diseases include deformation of legs and feet and destruction of the nasal and palatine bones. Diagnoses of treponemal diseases in skeletal remains are well established; however, clinicians and osteologists have mentioned that many other diseases may have the same symptoms as syphilis (Ferguson and Varnado 2006; Harper et al.

2011). Pinta does not leave any lesions on bone. In yaws and bejel, the tibia is the most infected bone (Buckley and Dias 2002). The overall change is a saber shin and an increase in the cortical diaphysis. In yaws and bejel, the skull, face, shafts of long bones, and hands are likely to be destroyed through bone erosion (Buckley and Dias 2002; Roberts 2000). In venereal syphilis, joints of the upper limb, knee, and ankle are likely to be infected as well as the skull, but the face is usually not affected in venereal syphilis (Buckley and Dias 2002; Roberts 2000). The skeletal manifestations of the long bones are similar to general periostitis or osteomyelitis, but it is their distribution that reveals the etiology. On the skull, the appearance is pitted and scarred and is called **caries sicca**. Even though fairly distinct skeletal lesions occur from treponemal infections, rates of skeletal effects are low; for endemic syphilis, 5% or less of infected individuals will have skeletal lesions, and for venereal syphilis, 10% to 20% of infected individuals have skeletal lesions (Shuler 2011).

Venereal Syphilis Origins

Although endemic treponemal infections seem to have been around since the origin of *Homo* (Meyer et al. 2002) and bioarchaeologists have found *Treponema* in both Old World and New World remains, questions of venereal syphilis origins remain. Smith (2006) discovered evidence of treponemal disease in subadults that date to around six thousand years ago in the Tennessee River Valley. Smith also found that around 1 BC, treponemal infections increased in frequency; 1 BC marks the onset of pottery and ceramics that led to increased sedentary behavior and, therefore, the possibility of population growth and crowded conditions. The afflicted individuals in the later population were adults, whereas in earlier populations those afflicted were mainly subadults, which suggested to Smith that in the later period more individuals survived their infections and there may have been more recuperative care.

Anthropologists are in disagreement over whether venereal syphilis was in the Old World prior to Columbus's return from the New World (see Harper et al. 2011; Heathcote et al. 1998; Merbs 1992; Rothschild et al. 2000). There are three main theories regarding the origin of syphilis: (1) Syphilis was in the Old World prior to Columbus's return, but it was mild and not widespread. Doctors could have misdiagnosed the disease, and there may have been a mutation that resulted in syphilis changing from the endemic version to the venereal version. (2) Syphilis was introduced to the Old World through Columbus's return. Studies of thousands of Old World skeletons have revealed no pre-1493 cases of syphilis in Europe, while historians and anthropologists have found evidence that in the 1500s, cases of syphilis suddenly appeared. Plus, the oldest case of syphilis is from around six thousand years ago in South America. (3) Multiple origins are possible, and *Treponema* was possibly present in both the New and Old Worlds, perhaps even originating from primates millions of years ago, but the environment dictates whether venereal or endemic syphilis will thrive. Evidence from the New World from Hutchinson and Richman (2006) may help to explain syphilis origins; they examined treponemal infections in the southeastern United States and found no evidence of venereal or congenital syphilis, but endemic syphilis was found in twenty-five sites dating from 8000 BC to AD 1600. New World syphilis has been around for a very long time, but venereal syphilis evidence is found only in the Old World, which could mean that syphilis mutated in the colder climes of Europe to spread more effectively in locations where the presence of clothing would minimize the transmission of syphilis through contact. However, Old World evidence of syphilis that dates prior to the return of Columbus has been found in the skeleton of a thirteenth-century Turkish individual; the skeleton belongs to a fifteen-year-old and has post-cranial lesions and destruction of the maxillary bone coupled with common dental anomalies associated with venereal syphilis, including Hutchinson incisors and malformed molars (Erdal 2006). Furthermore, Mays et al. (2012) examined a British skeleton from a female dating three hundred years before Columbus's return from the New World who seemed to have syphilis. It is possible that the individual actually had bejel, which was prevalent in Europe prior to the seventeenth century, or that the date of the skeleton is incorrect because the high marine diet of UK populations can alter the effectiveness of radioactive dates. In a meta-analysis, Harper et al. (2011) found that the data for Old World syphilis prior to Columbus's return is not convincing; the studies either lack specific diagnostic criteria or the dates are not firm. Some researchers have suggested that ancient DNA will resolve where venereal syphilis originated, but DNA research on syphilis has not been successful (Anastasiou and Mitchell 2013; von Hunnius et al. 2007). It appears that survival of syphilis DNA occurs in only very specific skeletal cases, such as in infant congenital syphilis (Anastasiou and Mitchell 2013).

Treponemal Infections Today

Treponemal diseases are still inflicting pain and disfigurement on individuals. In warm and humid environments, over-crowded urban areas where sanitation is low tend to breed environments ripe for yaws. Yaws is still endemic in fourteen countries. Yaws inflicts children most, and the highest rates are in those under fifteen years of age; 10% of people infected with yaws will be disfigured after five years. Yaws was the first disease to be slated for eradication after World War II (Rinaldi 2012). Yaws eradication was attempted between 1952 and 1964. According to the WHO, in the forty-six countries where eradication was attempted, the rate of yaws went from 50 million to 2.5 million, which is a 95% reduction (Rinaldi 2012). And in 1990, official reporting of yaws was discontinued. Currently, it appears that 2.5 million individuals have yaws and that there are nearly half a million new cases a year (Rinaldi 2012). No vaccine can help prevent yaws, but penicillin can be used to treat it, and with the oral treatment rather than the hypodermic treatment, eradication may still be possible (Rinaldi 2012).

Venereal syphilis affects some of the most disadvantaged in the United States; this is as true today as it was within the last one hundred years (de la Cova 2011; Ferguson and Varnado 2006). Although rates of infection have dropped since the 1990s, syphilis rates increased again in the twenty-first century (Ferguson and Varnado 2006). In the United States, syphilis rates are highest in blacks, males who have sex with males, and HIV-infected individuals (Ferguson and Varnado 2006). Even in the nineteenth-century remains from the Hamann-Todd, Terry, and Cobb skeletal collections, blacks had greater syphilis rates than whites, which have been attributed to crowded living conditions and poor sanitation (de la Cova 2011). However, even with the recent increase in syphilis, rates are still low in the United States, with the highest rate occurring in Georgia at 6.8 individuals per one hundred thousand (Ferguson and Varnado 2006).

MYCOBACTERIAL DISEASES

Tuberculosis and leprosy are the most frequently reported specific infections in the bioarchaeological literature (Roberts 2000). These infectious diseases are both caused by the genus *Mycobacteria*. Their rates increased through time from prehistory to the Middle Ages (Roberts 2000). Tuberculosis is caused by two infectious species: *Mycobacteria tuberculosis* and *Mycobacteria bovis*. Leprosy is caused by *M. leprae* or *M. lepromatosis*.

Tuberculosis

Tuberculosis (abbreviated as TB) is one of the leading causes of death today (Wilson 2012). The WHO estimates that half a million children get tuberculosis each year, and seventy thousand die from it (Burnwal et al. 2012). *Mycobacterium tuberculosis* destroys bone and hinders healthy new bone formation; the most common locations of lesions in skeletons are vertebrae, crania, and ribs. Tuberculosis prefers the red marrow of bones, where red and white blood cells along with platelets are produced. Red marrow is found in the thorax, which is why TB is usually found in vertebrae (Merbs 1992). Plus, the location of the bone reactions is probably because tuberculosis resides primarily in the lungs and the proximity of the lungs to the ribs. *Mycobacterium tuberculosis* is spread through the air by individuals with active infections whose symptoms include coughs. *M. bovis* is transmitted through non-pasteurized milk and through cattle. The hip and knee are common areas of infection (Roberts 2000). For both types of *Mycobacteria*, other symptoms include weight loss, fatigue, night sweats, painful breathing, lung damage, joint destruction, and vertebral collapse.

DIAGNOSING TUBERCULOSIS. Diagnosis of tuberculosis in the clinical setting may seem straightforward, but complications still arise. For instance, clinical tuberculosis cases are not easily diagnosed through bone scans (Cunha 2002; Santos and Roberts 2001). Since back pains are common, patients are not often tested for tuberculosis, and yet in one small study, ten out of fifteen individuals with cervical osteomyelitis had tuberculosis (Prasad et al. 2007). Another study found that in forty patients with vertebral tenderness, four types of infections were present: staph, strep, *E. coli*, and tuberculosis. And Colmenero et al. (2013) found that tuberculosis vertebral osteomyelitis was often detected late, and this resulted in poor prognoses. Burnwal et al. (2012) also showed that in children, 1% to 6% of tuberculosis osteomyelitis is not in the pulmonary region; the rate for non-pulmonary tuberculosis osteomyelitis is even higher in developing countries. Thus tuberculosis infection can be in unexpected locations, such as in the ulna in the case study presented by Burnwal et al. (2012). *Mycobacteria* can actually be found in any organ or bone as demonstrated by the five cases of tuberculosis of the sternum after heart surgery reported by Wang et al. (2007).

In the paleopathological record, many other diseases (such as pneumonia, brucellosis, or fungal infections) leave marks similar to TB on bones (Matos and Santos 2006; Mutolo et al. 2012; Roberts 2000). Lytic lesions of the spine are a good indicator of tuberculosis; these lesions make the bones look like Swiss cheese (Dabernat and Crubézy 2010; Merbs 1992). However, lytic lesions of the spine are rare in the remains of children's bones, and diagnosing children can be difficult. Dabernat and Crubézy (2010) found a young child dated to 3700 to 3200 BC from Egypt with thoracic vertebrae with lesions coupled with similar lesions on the clavicle, cervical vertebrae, scapula, and long bones. Parietal lesions can also be used to diagnose tuberculosis; Dawson and Robson Brown (2012) found a child with Swiss cheese–like holes in a UK site dated between AD 1150 and 1539. Some of the most effective ways for anthropologists to diagnose tuberculosis include examining ribs three through seven for bilateral bone remodeling and lesions. Right sides are more frequently affected than left sides. Nevertheless, some anthropologists ignore rib inflammation as an indicator of tuberculosis since clinicians do not include rib periostitis in their diagnostics for tuberculosis (Santos and Roberts 2001). Raff et al. (2006) have provided further support for rib examination as a diagnostic tool for tuberculosis identification with DNA tests on remains from the prehistoric Midwest United States dating between AD 1000 and 1200. Santos and Roberts (2001) argued that X-rays do not adequately reveal reactive rib changes, but the rib lesions are visible in macroscopic examinations. Thus clinical and bioarchaeological methods should not be identical. Additionally, collapsed vertebrae that result in a hunchback (also known as kyphosis or **Pott's disease**) are often associated with tuberculosis. In Mayan terra cotta pottery, tuberculosis vertebral kyphosis seems to be depicted (Mackowiak et al. 2005). Merbs (1992) commented that *M. bovis* can cause skeletal involvement beyond the pulmonary region. Plus, Kelley and El-Najjar (1980), using documented cases from the Hamann-Todd Collection, found that tuberculosis lesions varied greatly. Further complicating the diagnostics in skeletal remains is the fact that only between 3% and 10% of infected individuals experience skeletal changes (Shuler 2011). However, Anson et al. (2012) suggested that this low rate of skeletal changes in tuberculosis patients is in part due to the omission of rib examination in clinical studies.

TUBERCULOSIS ORIGINS AND SPREAD. Studies on the evolution of *Mycobacterium* species find that the disease complex evolved between thirty thousand and thirty-five thousand years ago (Dabernat and Crubézy 2010). Tuberculosis has been discovered in many skeletal remains from prehistory and historic times. Some early cases include a nearly 10,000-year-old Jordanian skeleton and a 5,400-year-old Egyptian mummy (Dabernat and Crubézy 2010). Skeletal remains from the Iron Age (around 400 to 230 BC) are the earliest evidence of tuberculosis in the United Kingdom; the remains had skeletal and DNA evidence of tuberculosis (Lewis 2011; Mays and Taylor 2003). Other early cases come from Hungary (Évinger et al. 2011), Chile (Mackowiak et al. 2005), Germany (Nicklisch et al. 2012), Siberia (Murphy et al. 2009), and Thailand (Tayles and Buckley 2004). Rates of tuberculosis seem to have increased drastically with the Middle Ages and the growth of European cities (e.g., Anderson and Carter 1995; Kjellström 2012).

Like syphilis, the origin of tuberculosis is also a controversial topic. One origin theory is that tuberculosis was brought over to the New World by Columbus because Native Americans seem particularly susceptible to the disease, but tuberculosis may have evolved in Asia and was brought over to the New World earlier (Daniel 2000; Merbs 1992). It has also been hypothesized that there may be multiple origins for tuberculosis. Some of the earliest reports of tuberculosis in the New World prior to European contact were published as early as 1893 (Merbs 1992). Lambert (2002) examined rib lesions in the remains of thirty-two individuals from Colorado and found that about a third of these pre-contact (AD 1075–1280) Amerindians had rib lesions that may have been the result of tuberculosis infections. More females and young individuals were afflicted, which is a pattern that is also found in modern clinical studies. These Colorado Amerindians occupied an area where competition for resources led to warfare, which may have increased stress and susceptibility to tuberculosis. Other cases of tuberculosis in pre-contact remains have been found in Arizona and New Mexico (Mackowiak et al. 2005). Yet these cases could also be related to other respiratory ailments that are common in sandy environments. More conclusive New

World evidence comes from Chilean mummies that show DNA evidence of tuberculosis (Mackowiak et al. 2005).

In the Old World, Suzuki and Inoue (2007) reported on cases of possible tuberculosis through collapsed vertebrae in the Yayoi of Japan who were sedentary agriculturalists. The Yayoi period (300 BC–AD 300) is marked by an increase in warfare, social stratification, and political stratification; it seems that immigrants in this area could have brought the disease with them, although Europeans had not yet made contact. Ancient DNA (aDNA) from Bornean skeletal remains and South American ice mummies provides additional evidence for pre-European tuberculosis in Asia and the New World. According to aDNA studies, tuberculosis probably has a zoonotic origin and may have arisen in any place where animals and humans were in close contact (Zink et al. 2007).

Due to the method of infection, tuberculosis rates seem particularly sensitive to population density, and thus it increased as urban areas flourished. Although tuberculosis has decreased in the developed nations, tuberculosis is still one of the top killers in the world today (Wilson 2012). The rate of tuberculosis in developed countries is about ten individuals per every hundred thousand, whereas in developing regions, such as Africa, Asia, and South America, country rates range from one hundred to two hundred individuals per every hundred thousand. Urban areas have higher rates of tuberculosis than rural areas; for example, in Pakistan, tuberculosis rates for urban areas are 329 per hundred thousand, whereas the rural rate is 171 per hundred thousand (Alirol et al. 2011). Additionally, rates of tuberculosis are closely correlated with poverty and malnutrition (Alirol et al. 2011; Roberts 2000). Immunosuppressed individuals, such as those with HIV, have higher rates of tuberculosis as well; Colmenero et al. (2013) found that nearly one in five individuals with tuberculosis vertebral osteomyelitis also had HIV.

Migration also leads to the spread of the disease; early migration and trade likely caused TB to become pandemic (Suzuki et al. 2008; Mackowiak et al. 2005). Travel to exotic places has become easier and cheaper in recent years; thus, more individuals from developed countries come into contact with pathogens that are foreign to them. For example, Waner et al. (2001) found that milk in rural Africa contains *M. bovis* and *Brucellosis*, and thus travelers to these locations who may have been told not to drink the water should also be wary of drinking unboiled milk.

However, even in the United States, tuberculosis is still problematic for at-risk populations. Ethnic differences in tuberculosis rates have been found in past and present US populations. For example, de la Cova (2011) examined 6,511 males born in the nineteenth century and found that African-Americans in autopsy collections had higher rates of tuberculosis than European-Americans.

In recent years, tuberculosis has also affected various parts of the US population more than other parts; homeless, poverty-stricken, and migrant or itinerant people are the most likely to have tuberculosis (Mayer 2000).

Antibiotic-resistant tuberculosis has arisen. Multiple-drug-resistant tuberculosis minimizes the effect of fighting the disease (Mayer 2000). In order to cure multiple-drug-resistant tuberculosis, a regimen of two to four medications a day for half a year is required; in New York City, only half of the patients engaged in treatment complete treatment (Mayer 2000). Although the eastern United States is most severely affected by drug-resistant forms, the ease of movement within the US has allowed the problem to expand westward (Mayer 2000).

Leprosy

Although leprosy (*M. leprae* and *M. lepromatosis*) is the result of a bacterium closely related to tuberculosis, most people have a natural immunity to leprosy. Yet there are geographic pockets (the tropics and subtropics of Africa, Asia, and South America) where leprosy thrives. Leprosy and tuberculosis often coexist and can be confused in the skeletal record. Leprosy is transmitted by inhalation or contact with an open wound (Kjellström 2012). Males are more often affected by leprosy (Kjellström 2012). Teens and reproductive-age females have milder forms of leprosy than most males (Guerra-Silveira and Abad-Franch 2013).

DIAGNOSING LEPROSY. The incubation period of leprosy averages between three and six years (Kjellström 2012). Its low virulence means that the disease is rarely fatal, but the more severe symptoms are correlated with early death (Boldsen 2001). Symptoms include loss of peripheral nerve sensation and poor blood circulation; minor cuts can lead to loss of fingers and toes since no healing occurs when the body is injured. The bacteria thrive in the extremities where they cause nerve damage (Kjellström 2012). Leprosy can lead to pyogenic osteomyelitis that is secondary to leprosy, which means that the osteomyelitis is caused by an additional bacterium. Or leprosy can cause osteomyelitis without pus creation (Andersen et al. 1994). Because leprosy causes peripheral nerve damage that is expressed in living people by paralysis of the extremities, individuals are prone to getting ulcers on their feet and not realizing the development of these lesions (Andersen et al. 1994). The ulcers then become infected and may result in erosion of the bone that fails to heal (Andersen et al. 1994).

Chronic infection from leprosy does lead to skeletal involvement that includes changes in the bones of the face, such as the erosion of the anterior nasal spine, and penciling (or distal thinning) of the hands and feet (Boldsen 2001; Merbs 1992; Shuler 2011). Bioarchaeologists examine skeletons for

symmetric and bilateral lesions of the lower legs, feet (especially the fifth metatarsal), and rhinomaxillary regions, including the nasal aperture, alveolar sockets, and palatine (Boldsen 2001; Kjellström 2012). However, other infections may mimic leprosy on skeletal remains; for instance, Blau (2001) reported on a possible case of leprosy dating from 2000 to 1200 BC in the United Arab Emirates; the **nasomaxillary** region was affected, but the author could not conclude whether this represented leprosy, treponemal disease, or a sinus infection. Additionally, leprosy rarely involves the skeleton; less than 5% of individuals with leprosy have skeletal lesions (Shuler 2011).

LEPROSY RISE AND FALL. The earliest known cases of leprosy occurred in Israel (1411–1314 BC), India (600 BC), Scotland (2300–2000 BC), and Egypt (200 BC) (Kjellström 2012; Larsen 1997). Written records about leprosy date back to 2000–1500 BC. Leprosy is first noted in Europe in the fourth century BC, and it became common in the Middle Ages in Europe; but then leprosy infections decreased once tuberculosis started to thrive (Kjellström 2012). When tuberculosis rose in frequency in Europe, leprosy rates decreased, perhaps as a result of interspecies competition, although leprosy and tuberculosis are often found in the same individual.

By AD 1000, leprosy reached Central Asia, perhaps as a result of trade with the Middle East. For example, Blau and Yagodin (2005) found evidence of nasal cavity resorption in West Central Asia that dates back to the third and fourth centuries AD. Leprosy evidence was found in pre-European-contact Micronesia (specifically Guam), which suggests that the bacteria may have spread from East Asian travelers (Trembly 1995). On the other hand, DNA studies show that leprosy in Europe and East Africa are similar and may have spread to Asia with trade (Wilson 2012). Evidence of leprosy in the New World prior to European contact is absent, which suggests that the origin of leprosy is in the Old World.

In Europe, leprosy was considered the dreaded disease of the Middle Ages, and individuals with leprosy were institutionalized and stigmatized (Boldsen 2001). Leprosy peaked in its frequency and severity in Europe from AD 1400 to 1500 and disappeared by the 1900s, but there were a few surges in Iceland and Norway (Kjellström 2012). A possible case of leprosy in a young adult male in nineteenth-century England was discovered by Walker (2009). By this time, leprosy infections had mostly disappeared from Europe; thus, Walker suggested that the young male may have been to India or another British colony. The afflicted male's treatment included amputation, and he died of another infection.

Leprosy eradication has led to a 90% decrease in leprosy in the last two decades; but new cases do appear, and more recently no change in rates has occurred (Dogra et al. 2013). In the United States and other developed nations, leprosy is very rare (Wilson 2012). It appears that the extreme disfiguring and institutionalization of those afflicted, coupled with aggressive treatment of the disease, has led to the overall failure of leprosy to become pandemic. Leprosy does still occur in various pockets of the world. There are one hundred thousand new cases a year, many of which are in India, but there are also many new cases in Brazil each year, with estimates ranging up to forty thousand a year (Dogra et al. 2013). Brazil's cases seem to stem from tattoos, dog bites, and armadillos. However, not all researchers agree that armadillos pose a threat to humans (Schmitt et al. 2010).

CONCLUSIONS

Even in today's world, infectious diseases are the leading cause of death. The problem is threefold according to the NIH (2013): (1) there are new diseases emerging, (2) old diseases are reemerging, and (3) some diseases have been thriving persistently. Developed nations, for example, have seen a recent increase in some diseases as a result of the decrease in vaccination; measles and whooping cough are making a resurgence in Europe and the United States. Polio is another disease that threatens to reemerge. By 1979, the United States was declared polio-free. However, in countries throughout the developing world, such as Afghanistan, Nigeria, and Pakistan, polio still thrives (Gholipour 2013). Recent outbreaks in Syria and China have led the WHO to raise concern that polio may be reintroduced to countries that have been declared polio free (Gholipour 2013). With vaccination levels decreasing, polio's reintroduction into Europe may have dire consequences; the WHO is thus attempting a complete eradication of polio by 2018. Some chronic diseases are now attributed to infectious diseases, which emphasizes the importance of preventing and curing bacterial, viral, fungal, and protozoan infections. For example, chronic gastric ulcers, which used to be linked to diet and stress, are now linked to the common pathogen *H. pylori*.

Globalization through travel and trade has led to the spread of various infectious diseases. Some of these may affect small groups of people, such as the diseases spread by exotic pets; 3% to 5% of *Salmonella* cases are related to pets (Kolker et al. 2012). And according to the NIH (2013), exotic rodents, for example, have brought over monkey pox to the United States. Sushi parasites, which used to be mostly confined to Japan, are now spreading all over as a result of shared cultures (Bucci et al. 2013). Other forms of globalization have widespread consequences. For instance, there are one million international travelers daily, and these travelers may bring endemic pathogens into new environments (Mayer 2000). A form of antibiotic-resistant gonorrhea once found only in Southeast Asia, for instance, has been found in Washington, D.C. The global market has created jobs in urban areas and increased movement to cities; in 2009, half of the world's population lived in urban areas (Alirol et al. 2011). In developed countries, urban areas increase health, but in developing countries, it is the opposite (Alirol et al. 2011). For example, in Sudan, 94% of urban dwellers live in dismal shantytowns with low sanitation, inadequate sewer systems, and high rates of infectious diseases that lead to morbidity and mortality (Alirol et al. 2011; Waldvogel 2004). Urban areas in developing countries have led to a resurgence of tuberculosis, cholera, typhoid, plague, and yellow fever (Heymann and Rodier 2001). One of the most important problems in infectious disease studies includes antibiotic-resistant forms of bacteria.

In 1969, the medical community declared that they had won the war on infectious diseases (Mayer 2000). Yet the emergence of the AIDS pandemic revealed that the war was still going strong (Mayer 2000). Also, in the 1990s, the CDC noted that infectious diseases were a major health concern again, and in 1995, stories about *E. coli*, *Hantaan* virus, and *Cryptoporidiosis* made the news (Mayer 2000). Antibiotic-resistant bacteria in the forms of *Staphylococcus*, tuberculosis, flu, gonorrhea, and *Candida* have increased with the overuse and misuse of antibiotics. According to the NIH (2013), between 5% and 10% of hospital patients develop infections that are antibiotic resistant. In 1992, about 13,300 patients died from infections in US hospitals; in 2012, the number has gone up to 90,000. In the Soviet Union, multiple-drug-resistant tuberculosis occurs in 20% of all cases (Heymann and Rodier 2001). Although many infections are antibiotic resistant, most of these do not affect the bones and thus will leave no skeletal record for future populations. Dhanoa et al. (2012) reported that only 1% of all community-acquired methicillin-resistant *Staphylococcus aureus* causes osteomyelitis. It is commonly thought that antibiotic overuse and misuse is the cause of resistant forms, but even correct use will result in resistant forms, albeit at a slower pace (Spellberg et al. 2013). Bacteria found in a cave that was underground for four million years was found to be antibiotic resistant; evolution occurs regardless of whether humans drive it or not (Spellberg et al. 2013). To slow the evolution of antibiotic-resistant bacteria, narrow drug use (just for specific diseases) and eliminating antibiotics from the environment, such as in the soil and in foods, may help (Spellberg et al. 2013). Determining just the right length of time to take antibiotics for an infection is essential to slowing down the evolution of antibiotic-resistant forms of bacteria (Spellberg and Lipsky 2012).

CHAPTER 9

Congenital Defects

ACCORDING TO THE WHO (2013), ONE IN THIRTY-THREE INFANTS IS BORN with birth defects, which are also referred to as congenital anomalies or congenital defects. Over a quarter of a million individuals die during their first month of life every year from birth defects; the NIH (2013) estimates that thousands of different birth defects exist and that they are the leading cause of death in the first year of life in the United States.

According to the WHO (2013) and the March of Dimes (2013), the most common birth defects are heart defects, **neural tube defects** (which include spina bifida), **cleft palate**, and **Down syndrome**. Some other birth defects include Marfan syndrome, which may include kyphosis and a protruding sternum; fragile X syndrome, which may include a narrow face and a prominent forehead and chin and is the most common inherited intellectual disability in the United States; **clubfoot**, which is often found in conjunction with cerebral palsy and spina bifida; and various forms of **dwarfism**. Birth defects can be split into structural, which involve a problem with a specific body part, or functional, which include problems with the functioning of body parts. For example, cleft palate is a structural birth defect, whereas autism is a functional birth defect. Only 15% to 25% of all birth defects can be attributed to specific genes, and somewhere between 65% and 75% of birth defects are of an unknown etiology (Ndreu 2006). Research suggests that only 10% of birth defects are attributable to the environment, and less than 1% of birth defects are attributable to prescription medicines, chemicals, radiation, and hyperthermia (Ndreu 2006). For most birth defects, etiologies are complex and may be genetic, environmental, or even related to infections. Adequate vitamin B-9 (which is known as folate when it occurs naturally and folic acid when it is man-made and used in food fortifications), vaccinations, and prenatal care can reduce the risk of birth defects. Malnutrition, alcohol consumption, and parental age all contribute to birth defect risks. Nearly all birth defects occur in low- and middle-income countries.

BIRTH DEFECTS DIAGNOSTICS

I will only address birth defects visible in skeletal remains. For the most part, macroscopic examination of the patient or the skeletal remains determines the diagnosis. However, since many birth defects have similar and overlapping traits, determining specific causes of the defects is difficult. This is especially true in bioarchaeology where trauma, trephination, infections, and even postmortem damage may mimic some birth defects; for example, Blau (2005) wrote an article about a child from West Central Asia with an **aperture** in the skull that may relate to trephination surgery, pathologies such as infections, or congenital causes. Yet anthropometry, which is the use of measurements taken on living patients based on anthropological methods, has proved useful in diagnosing diseases, such as Huntington's disease (Farrer and Meaney 1985) and Treacher Collins syndrome (Kolar et al. 1987), on skeletal remains. Conversely, use of clinical diagnostics is difficult on skeletal remains since many diagnoses include data on soft tissue, behavior, pain, movement, and blood work. Furthermore, the use of X-rays and MRIs, rather than direct bone examination, may cause doctors to miss some subtle variations in the patient (Mulhern and Wilczak 2012).

Birth Defects in the Osteological Record

Nearly all examples of birth defects recorded in the bioarchaeological record are case studies; anthropologists find skeletons that look different from normal or expected, and then they compare the traits with known pathologies. Cope and Dupras's (2011) case study of osteogenesis imperfecta in an Egyptian skeleton and Roberts et al.'s (2004) case study of a Romano-British fourth-century male with a clubfoot are good examples. The articles review other possible diagnoses and summarize the other possible similar cases reported on. The types of birth defects found in the bioarchaeological record include vertebral anomalies (such as spina bifida), other neural tube defects (such as hydrocephaly), missing bones, aural atresia, Klippel-Feil syndrome, osteogenesis imperfecta, and dwarfism.

VERTEBRAL ANOMALIES. Vertebral anomalies are the most frequent types of congenital anomalies reported in the bioarchaeological record; spina bifida, **sagittal clefting**, scoliosis, and other variants are reported in a variety of collections. For example, Keenleyside (2012) found sagittal clefting of vertebral bodies that resulted from failure of the two lateral **chondrification centers** to fuse during development. This anomaly usually occurs after three to six weeks of **gestation**. Keenleyside reported on a Greek female between twenty-five to thirty years of age dating to 610 BC who had sagittal clefting; she also reported that sagittal clefting often occurs with other anomalies, such as kyphosis or scoliosis. In living populations, sagittal clefting has been associated with back pain, but it can also be asymptomatic, which makes determining rates of occurrence difficult. Other bioarchaeological examples of sagittal clefting have been found in Peru, the United States, Canada, Siberia, the Arctic, and the United Kingdom (Anderson 2003; Keenleyside 2012). Pitre and Lovell (2010) reported on several vertebral anomalies in a female in her forties from historic Quakers living in Ontario between 1663 and 1814. The skeletal remains exhibited abnormal sacral shape and scoliosis. These variations are often congenital but can be influenced by environmental factors rather than genetics. For instance, low vitamin B-9 or zinc can result in vertebral anomalies. These pathologies are also found with clubfeet, fused ribs, and defects of the kidney and heart.

One of the most common types of vertebral anomalies is spina bifida, which results from a lack of fusion of the neural arch of the spine. Spina bifida accounts for half of the three thousand neural tube defects in the United States each year (March of Dimes 2013). The most severe form of spina bifida includes **anencephaly**, which will result in death at birth or shortly afterward (Jorde et al. 1983). Neural tube defects are more common when the mothers are obese, diabetic, or taking anti-seizure medicines. Folate, which is a vitamin B that is found in legumes, reduces the risk of spina bifida (and other neural tube defects). Inbreeding may increase the risk of spina bifida. In the archaeological record, the least severe form of spina bifida—spina bifida occulta—is often written about, but the medical community does not even count it as a birth defect. In living people, the appearance of a dimple or hair spot over the location may reveal spina bifida occulta. Figure 9.1 displays clear evidence of spina bifida occulta. Figure 9.2 shows evidence of a thoracic vertebra with an unfused arch. El-Din and El Banna (2006) examined congenital anomalies of the spine and found that ancient Egyptians had spina bifida occulta at a rate of about 3%, but modern rates in Egypt are difficult to assess. Hussien et al. (2009) found that spina bifida occulta occurred in six out of ten individuals in a prehistoric oasis, which may relate to inbreeding, whereas in the larger city of Giza, Egypt, the rate was 3%. Inbreeding has also been blamed for high rates of spina bifida in Oregon/California border proto-historic Indians (Bennett 1972). The few studies on more recent samples include Willis's (1923) and Eubanks and Cheruvu's (2009) studies of osteological specimens from the Hamann-Todd Collection. Willis noted that spina bifida occurred mainly in the lower back and that it was present in males only and in more whites than blacks. He noted that the frequency of spina bifida in the Hamann-Todd Collection was 1.2%. Eubanks and Cheruvu also found that males were more frequently affected than females and that white rates of spina bifida were higher than black rates, which concurs with the Willis study. But, unlike Willis, Eubanks and Cheruvu discovered 355 individuals out of 3,100 with spina bifida occulta, which amounts to 12.4%—ten times the rate that Willis

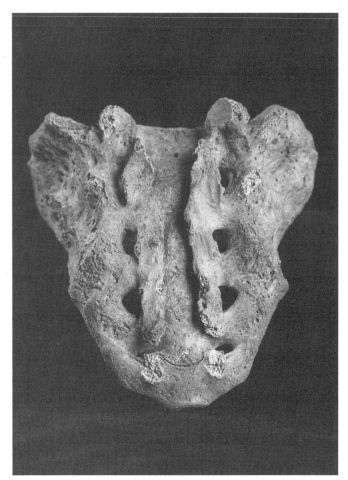

Figure 9.1. Spina bifida occulta. The least severe form of spina bifida only involves the sacrum, such as in this twenty-six- to thirty-five-year-old male. In the center of the sacrum, the left and right side fail to come together to form an arch. Photograph by Daniel Salcedo.

reported in 1923. The ethnic difference in spina bifida may be attributed to folate protection that dark skin bestows. Folate is destroyed by UV rays, and the high density of melanin in dark-skinned individuals prevents UV rays from over-penetrating the skin and thereby protects folate. More severe forms of spina bifida are spina bifida cystic and aperta; these appear to be different expressions of the same gene (Masnicová and Beňus 2003). In a rare case of more severe spina bifida, Dickel and Doran (1989) found a fifteen-year-old skeleton dated to about 7,500 years ago with spina bifida aperta that would have caused paralysis (as was evidenced by disuse **atrophy**).

Curate (2008) reported on a rare abnormality of a separated odontoid process of the second cervical vertebra found in an 840 BC adult Portuguese

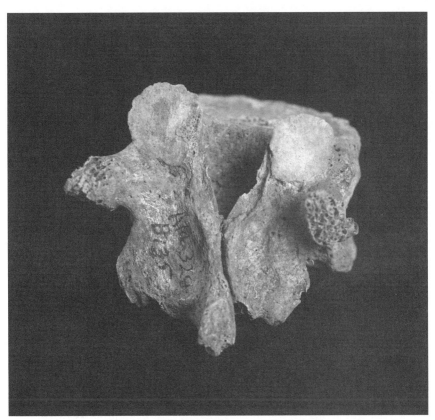

Figure 9.2. Unfused thoracic vertebral arch. Found in a thirty-nine- to forty-four-year-old male, this is a more severe case of a neural tube defect than usually seen in the archaeological record, but it is still a minor form of spina bifida and would not likely be noted in medical research. Photograph by Daniel Salcedo.

female. Curate proposed that the etiology can be either acquired or congenital; if it is congenital, then it likely is a result of fusion failure and may be associated with trisomy 21 (Down syndrome) or Klippel-Feil syndrome. Klippel-Feil syndrome is the fusion of cervical vertebrae that results in a short neck, low posterior hair line, and restricted mobility of the upper spine (NIH 2013). It is often associated with cleft palates, kidney problems, and heart malformations. Physical features include abnormal heads, faces, and sex organs. Pany and Teschler-Nicola (2007) described an AD 980–1018 teenage male with Klippel-Feil syndrome along with spina bifida occulta. Oxenham et al. (2009) also reported on a male with Klippel-Feil syndrome; this 3,500-year-old North Vietnamese individual was also likely paralyzed as evidenced by limb atrophy.

CLEFT PALATE. Cleft palate is among the most common forms of developmental malformation reported at birth (Phillips and Sivilich 2006). It occurs in about one in a thousand births. Cleft palate occurs when there is incomplete fusion of the mesenchymal tissue of the palatine process and bones. Fusion should occur during the first trimester. Much variation occurs in cleft palates, and **supernumerary** (or extra) **teeth** may be present; two examples of cleft palate are provided in figures 9.3 and 9.4. Cleft palate is more common in females than in males. The etiology of cleft palate is complex; there are around four hundred genetic and environmental causes of cleft palate (March of Dimes 2013). For example, tobacco use, folate deficiency, and maternal rubella can all cause cleft palates. Cleft palate may also be inherited. Asians and Native Americans are more prone to cleft palates than other populations (March of Dimes 2013). Cleft palates also occur in 150 syndromes (Phillips and Sivilich 2006). Symptoms include problems with hearing, speech, and breastfeeding (Phillips and Sivilich 2006). Cleft palates are rare in the archaeological record because these deformities likely led to mortality in most infants, and infant bones do not preserve well. Nonetheless, there is prehistoric archaeological evidence of cleft palates from California (e.g., Brooks and Hohenthal 1963; Weiss 2008) and Indiana (Phillips and Sivilich 2006). Currently in developed nations, most individuals are operated on shortly after birth.

SKULL ANOMALIES. In the bioarchaeological record, there are few cases of skull abnormalities. Usually these neural tube defects would result in early death and thus do not result in preservation. However, several case studies have been published on cranial birth defects. For example, Duncan and Stojanowski (2008) described a case of **craniosynostosis** in a sixteenth-century skull from the Southeast United States. Craniosynostosis is the premature closure of cranial sutures that results in seizures, blindness, and retardation according to the NIH (2013). The physical result on the skull is an irregularly long and small skull as illustrated in figure 9.5.

Sulosky Weaver and Wilson (2012) described a variety of congenital herniations of the cranium that have been studied from cases in the United States, Australia, Peru, Mexico, and Uzbekistan. Furthermore, they reported on a case of the least severe form of cranial herniations, atretic cephalocele (which involves the protrusion of the meninges), in an adult female from Sicily dated to about AD 630. These hernias, which at their most severe can expose part of the brain, result in susceptibility to trauma and infections (Sulosky Weaver and Wilson 2012).

Another form of cranial defect, which can actually be congenital or the result of infection or trauma, is **hydrocephaly**. Hydrocephaly is the abnormal accumulation of cerebrospinal fluid that usually results in death in the first eighteen months of life. Murphy (1996) described a case of what was likely congenital hydrocephaly in a six- to seven-year-old from medieval Northern Ireland; there are likely thirty cases of hydrocephaly in the archaeological

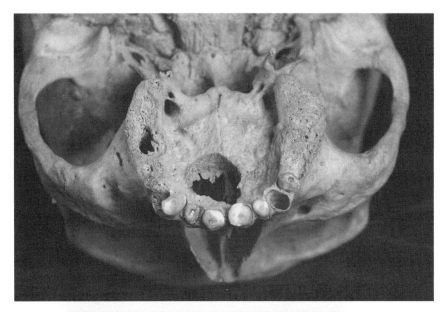

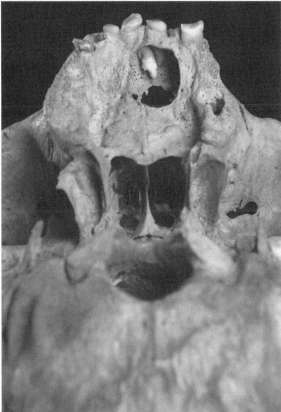

Figure 9.3. Cleft palate. The hole in this maxilla may be mistaken for an abscess (top), but upon closer inspection one can see the supernumerary teeth (bottom) that are common in cleft palates. The loss of teeth and bone resorption may be related or independent of the cleft palate condition in this elderly female. Photographs by Daniel Salcedo.

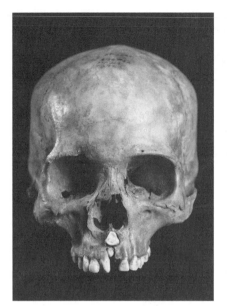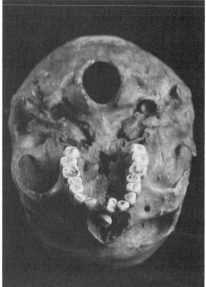

Figure 9.4. Cleft palate. Another example of a cleft palate from frontal (left) and inferior (right) views; this twenty- to twenty-four-year-old female had supernumerary teeth and bony reaction on her face as well. Photographs by Daniel Salcedo.

record, including a 2500 BC to AD 500 California individual who was likely paralyzed (Richards and Antón 1991).

Aural **atresia** is also found in the archaeological record. Aural atresia is the absence of an **external auditory meatus**; it can be congenital, infection related, or trauma induced (Masnicová and Beňus 2001; Swanston et al. 2011). Congenital atresia occurs in one in ten thousand to twenty thousand births; bilateral aural atresia, which would result in complete hearing loss, is rarer than unilateral, which results in hearing loss on just one side. Aural atresia, which was reported in a late nineteenth-century Canadian (Swanston et al. 2011), in a fifty-year-old female from fifteenth- to eighteenth-century Slovakia (Masnicová and Beňus 2001), and in an adult female from prehistoric Iowa (Hodges et al. 1990), can be associated with syndromes such as Treacher Collins, Crouzon, Turner, and Goldenhar. Aural atresia is sometimes found with cleft palates too (Hodges et al. 1990).

MISSING OR EXTRA BONES. Congenital absences of bones, such as a missing ulna, or fusions of various bones, such as fusion of the humerus and radius, are extremely rare. In the clinical records, absence of the proximal portion of a limb is called **phocomelia**. Meromelia is the absence of a particular part of a limb. Hemimelia is the absence of the extremities. And **amelia** is the absence of entire limbs. There appear to be no trends in limb absences and

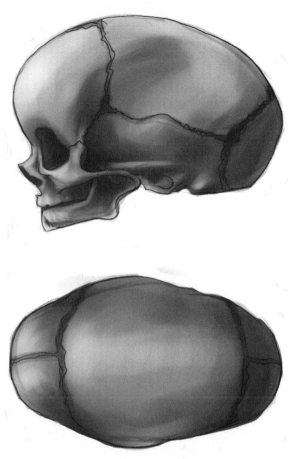

Figure 9.5. Craniosynostosis. An illustration of the abnormally shaped skull as a result of craniosynostosis. Illustration by Vanessa Corrales.

fusions; they are rare, and all varieties combined occur only in about five for every ten thousand births. The drug Thalidomide caused a brief epidemic of phocomelia and amelia in Europe, Australia, and Japan; it was not approved in the United States. Thalidomide came out in the late 1950s and was prescribed as an anti–morning sickness drug; shortly after the drug become popular, ten thousand children were born with limb defects (Kim and Scialli 2011). By 1961, the evidence that Thalidomide causes limb defects was strong enough that most countries banned its use. Currently, Thalidomide is back on the market, but for treatment of leprosy, tumors, and HIV wasting (Kim and Scialli 2011). Although the control of Thalidomide in the United States and other developed countries will likely prevent birth defects, prescriptions in developing countries (especially those with leprosy endemics) may lead to a new surge in limb birth defects (Kim and Scialli 2011). Usually limb reductions or absences are not found in the archaeological record, which may be a result of destruction of the bones after death or even infanticide. However, there are cases that have been found, such as in a Native American from Alabama dated to AD 1200 to 1400 with a lack of an ulna and a humeroradial synostosis (Mann et al. 1998). Similar cases from California and New Mexico were reported by Antón and Polidoro (2000). Gładykowska-Rzeczycka and Mazurek (2009) reported on an eighteenth-century Polish case with hemimelia in a fifty-five- to sixty-year-old female.

Extra fingers (**polydactyly**) and supernumerary teeth have also been found in the bioarchaeological record. Figure 9.6 is an illustration of one variation

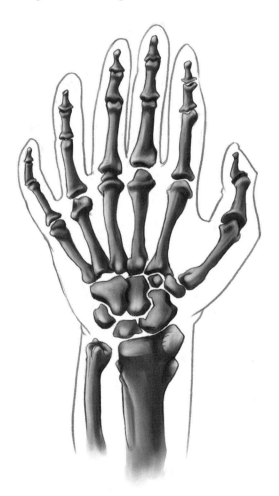

Figure 9.6. Polydactyly. An example of a common form of an extra finger found often in inbred populations. Illustration by Vanessa Corrales.

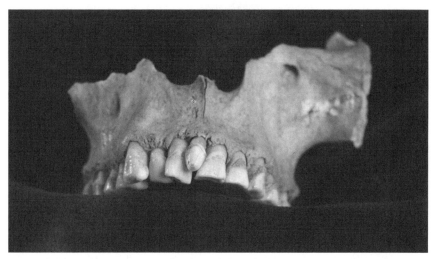

Figure 9.7. **Supernumerary teeth.** This adult male has an extra incisor that is peg shaped. Photograph by Daniel Salcedo.

of polydactyly seen in human populations with inbreeding. Figure 9.7 shows an individual with an extra incisor. Case et al. (2006) examined polydactyly in two individuals from the prehistoric American Southwest. They reported that a Y-shaped fifth metatarsal or metacarpal is often a trait of polydactyly; usually the extra finger or toe is after the last digit or before the first digit.

DWARFISM. Dwarfism is rare and occurs in one in fifteen to forty thousand births. There are various forms of dwarfism, such as anchrondoplasic, hypothyroid, and chrondoplasic; some forms of dwarfism are extremely rare, occurring in just two out of over two million births (Arcini and Frölund 1996). Genetic mutations, both heritable and random, are the cause of the different types of dwarfism. In most cases, dwarfs can pass on their birth defect; according to the March of Dimes (2013), if two adult dwarfs have offspring, there is a 25% chance of having a normal baby. Although dwarfism is rare, there are several cases in the bioarchaeological record. Two of the oldest cases of dwarfism are from a twenty-four-thousand-year-old Russian site; it appears that the individuals had deformity of the femora, short stature, and anteroposterior bowing (Formicola and Buzhilova 2004). Slon et al. (2011) reported on an Israeli find from the fifth to eighth century that appeared to be an adult male dwarf; Arcini and Frölund (1996) found two dwarfs in eleventh- to thirteenth-century Sweden, which they think were siblings; Frayer et al. (1988) reported on a dwarf from Italy; and Kozieradzka-Ogunmakin (2011) reported on an adult male dwarf dated to nearly 3000 BC from Egypt. Hernandez (2013) described a six-thousand-year-old case of a hypothyroid dwarf from China.

TEMPORAL CHANGES IN BIRTH DEFECTS

Although many environmental and genetic factors influence birth defect risk, the topics discussed below were chosen because they affect large numbers of people, have shown recent trends, and relate to structural birth defects that can be seen in the skeletal record.

Toxins

The evidence of pollution's effect on birth defects is limited. Yet there is some evidence that lead, pesticides, and chlorine may lead to birth defects. Pesticides used in the home or workplace may have negative effects on infant health. For example, using a sample of 156 Mexican-Americans, Brender et al. (2010) found evidence of a 5.6 times greater risk of neural tube defects with bombs or foggers (for pest control) in the home. They also found that females who walked through fields treated with pesticides had an increased risk of having offspring with spina bifida. Sherman (1996) reported on four children (two were siblings) with cleft palates, hydrocephaly, and microcephaly that had been exposed to flea control treatments that contain **chlorpyrifos**. And in South Africa, the reuse of plastic bottles that held pesticides has been linked to an increase in birth defects (Hereen et al. 2003). Industrial pesticides used for agriculture have been tied to an increase in birth defects, such as neural tube defects and limb reductions (Ochoa-Acuña and Carbajo 2009).

Other toxins include those that come from industry. Lead from ceramic factories in northern Italy, for example, was found to cause an increase in cleft palates in the region (Vinceti et al. 2001). And a chemical spill in 1979 from an IBM factory in New York led to an increase in low birth weights (Forand et al. 2012). Yet data from an industrial park with seventeen factories in Israel found that there was no increase in birth defects in the region in the Jewish population (Sarov et al. 2008). In China, arsenic levels in water from industry were inversely related to birth defects, which Wu et al. (2011) thought related to arsenic awareness and that people were careful in high-arsenic areas, but not so careful in low-arsenic areas. Also in China, chlorinated water as a by-product of industry has shown no consistent effects on birth defect rates (Hwang and Jaakkola 2003).

Nuclear plants have been a concern in many places, such as Japan, the United States, and Europe. The Chernobyl disaster in 1986 made many aware of the dangers of nuclear energy. Birth defects included spina bifida, severe limb reductions and deformations, hydrocephaly, and microcephaly. The birth defect rate in the affected region increased from 1.8% to 5% (Kozenko and Chudley 2010). Right after the accident, an increase in terminations of pregnancy and a reduction in conceived pregnancies may have resulted in fewer than expected birth defects (Kozenko and Chudley 2010). The closure of nuclear reactors in the United States has led to a decrease in congenital anomalies in infants who were born or conceived downwind of the reactors (Mangano et al. 2002). Even though the impression of birth defects caused by environmental pollutants and toxins is dramatic, most birth defects are not caused by these factors.

Consanguineous Marriages: Inbreeding Depression

Consanguineous marriage is a marriage between two closely related people. The most common form of consanguinity is cousin marriages, but uncle-niece and even sibling marriages occur. In the West, consanguinity is often regarded with aversion; inbreeding is not a part of American or European culture at present. Conversely, in parts of Asia and the Middle East, consanguinity is considered preferable; rates of consanguineous marriages range from 20% to 70%; most of these are first-cousin marriages (Hamamy et al. 2005). In India, consanguinity occurs at about 40% in both the South Indians and the Muslims in the north (Ramegowda and Ramachandra 2006). Consanguineous marriages are also common in North Africa and West Asia (Bittles and Black 2010). Since many consanguineous marriages occur in Muslim countries, people may erroneously think that Islam prescribes relative marriages, but it appears that in these regions non-Muslims also engage in consanguinity (Teebi and El-Shanti 2006). Consanguinity is likely tied to tribal societies as a way to pool resources (Bittles and Black 2010).

Although some anthropologists have tried to reduce the stigma against consanguinity (Bittles and Black 2010), other researchers have highlighted the negative aspects of consanguinity that include an increased risk of **autosomal recessive** birth defects, such as congenital heart disease (e.g., Ramegowda and Ramachandra 2006), primary congenital hydrocephaly (e.g., Ten Kate et al. 2013), cleft palates (e.g., Shawky and Sadik 2011), and foot and hand anomalies (e.g., Bromiker et al. 2004; Yüksel et al. 2009). In skeletal samples, increased cranial defects in Switzerland (Kutterer and Alt 2008) and higher dental arch asymmetry in Japan (Schaefer et al. 2006) have been linked to consanguinity.

Clinical studies using large samples have found that consanguinity also has negative health consequences. Some studies rate general health through anthropometric measures, such as height and head circumference. In a French study by Schreider (1967), inbreeding was found to negatively correlate with height, especially during the teen and adult years, and inbreeding was also found to negatively correlate with head size in a sample from Delhi (Krishan 1986). Not all studies find inbreeding depression with general anthropometric measures, such as head circumference and height (e.g., Khlat 1989; O'Brien et al. 1988). Studies on birth defects provide stronger evidence on inbreeding depression.

Many studies have focused on the rates of birth defects and infant mortality in consanguineous relationships. The hypothesis is that harmful recessive alleles that code for birth defects will be more commonly inherited in the homozygous pattern and therefore the defect will be expressed more frequently in inbred populations than in outbred (**panmixia**) populations. Using a large sample of 13,582 medical charts, Shawky and Sadik (2011), for example, found that consanguinity increased rates of structural congenital anomalies, such cleft palate and Down syndrome. They also reported that structural congenital anomalies correlated with a decrease in head circumference and decreased intellectual abilities. Bromiker et al. (2004), examining congenital malformation data from four hospitals in West Jerusalem between 1998 and 1999, found that first-cousin marriages had a rate of birth defects of 8.7%, and even in second-cousin marriages the birth defect rate was 6%; non-related parents had a birth defect rate of 2.6%. Some of the malformations that were found to increase with consanguinity included clubfoot, cleft palate, and Down syndrome. In Turkey, a study of over four hundred patients found that the rates of birth defects (such as Down syndrome, hand and foot anomalies, and cleft palate) were as high as 8.2% in consanguineous marriages, whereas in non-related mates the rates were as low as 1.4%. In the United Kingdom, Pakistani, Indian, and Bangladeshi mothers had higher rates of congenital malformations, which may have been related to consanguinity; whereas a population with a highly mixed gene pool, the West Indies, had lower rates of congenital malformations (Balarajan and McDowall 1985). A large study in Strasbourg, France, also revealed a relationship between congenital hydrocephaly and consanguinity (Stoll et al. 1992). In North America, religious sects, such as the Amish, have high rates of consanguinity; brother-sister and uncle-niece relationships are common among the Amish (Dorsten et al. 1996). Ellis–van Crevald syndrome, which is a form of dwarfism, has been linked to a recessive gene that is more commonly expressed in the Amish. Additionally, infant mortality is increased in the Amish consanguineous pairings (Dorsten et al. 1996).

As the world becomes more globalized and more urban, and as people's wealth is less dependent on land and other material resources and more dependent on wage labor, consanguinity seems to wane. Hamamy et al. (2005) found that consanguineous marriages decreased in Jordan over the last fifty years. Furthermore, they suggested that although campaigns to warn of consanguinity and birth defects have not worked, education—especially of females—has led to a decrease in consanguinity.

Parental Age, Screening, and Fertility Treatments

Increased female education and a greater number of females entering the workforce have played a role in delayed childbearing. For example, in British Columbia in 1971, 4.3% of mothers were thirty-five years old or older, while in 1988 that rate grew to 6.8% (Baird and Sadovnick 1991). In the

United States, 12% of mothers in 1986 were in their early thirties, whereas ten years earlier the rate was 3% (Baird and Sadovnick 1991). From 1980 to 2003, there has been a 70% increase in mothers over thirty-four years of age in the United States; at the same time, mothers under thirty years of age have decreased in frequency (O'Leary et al. 2007). The phenomenon is not just prevalent in North America; in Taiwan from 1999 to 2009, the rate of mothers over thirty-four years old grew from 4.3% to 9.4%. As people put off having children to later ages, birth defect rates should increase. Some results from national and state databases using large sample sizes have found that birth defect rates are complicated by screening, decisions to terminate births, and artificial reproductive strategies.

It is well known that chromosomal defects increase with maternal age. At the age of thirty-five, the risk of a chromosomal abnormality is 1 out of 192 births, but at age forty it is 1 out of 66 births according to the March of Dimes (2013). Down syndrome increases from 1 out of 1,250 in twenty-five-year-olds to 1 out of 30 in forty-five-year-olds. Khoshnood et al. (2000) examined Down syndrome in Mexican-Americans, African-Americans, and European-Americans using the National Center for Health Statistics data that were collected between 1989 and 1991 and found that in all groups, maternal age increased the risk of Down syndrome, but that the rate was higher in Mexican-Americans and African-Americans. The authors attributed these differences to access to screening; that is, the differences in Down syndrome rates are due to a higher rate of pregnancy termination among European-Americans compared to Mexican-Americans and African-Americans. Collins et al. (2008) used a birth defects registry database with data from 1986 to 2004 and found that in Australia, Down syndrome birthrates did not increase even with greater maternal ages. The lack of an increase in Down syndrome is likely the result of an increase in early-term testing, which led to an increase in pregnancy termination; in Australia, as in the United Kingdom, only 5% of future mothers who receive news that their child will be born with Down syndrome choose not to terminate the pregnancy (Collins et al. 2008). This is not always the case; in Malaysia, for instance, this same trend has not been found, perhaps because screening is less accurate (Ho et al. 2006).

When one discusses the impact of older mothers on offspring health, older fathers' impacts should also be considered. Studies have suggested that fathers over forty years of age may increase the risk of cleft palate, limb deformities, and hydrocephaly (see references within Kazaura et al. 2004). Yet maternal and paternal ages are highly correlated, and thus maternal age must be controlled for. Using a large national medical registry database from Norway, Kazaura et al. (2004) found that older fathers caused no increased risk in birth defects. Kazaura and Lie (2002) also found that Down syndrome was only associated with the mother's age. However, studies from a national registry in Poland provided evidence that cleft palate increased with the age of the father independently of the mother's age (Materna-Kiryluk et al. 2009). From the California Birth Defects Monitoring Program, Grewal et al. (2012) found that paternal age was associated with limb anomalies; they suggested that this relates to the genetic integrity of sperm.

Many older couples have difficulty achieving parenthood naturally and turn to artificial reproductive technologies (ART), such as in vitro fertilization (IVF) and IVF with intracytoplasmic sperm injection (IVF-ICSI). Williams et al. (2010) reported that between 1992 and 2006, offspring resulting from ART increased from 0.5% to 1.7% in the United Kingdom. Kelley-Quon et al. (2013) reported that in the United States, sixty thousand infants a year are born with the help of ART. As far back as 1987, reports of increased neural tube defects in ART offspring arose (Williams et al. 2010). Artificial reproductive technologies have also been linked to an increase in birth defects in Australia and Sweden (Kelley-Quon et al. 2013). Major congenital anomalies are 30% higher in infants conceived through ART than in natural conception; spina bifida is five times more likely in ART-conceived offspring (Williams et al. 2010). With IVF-ICSI, congenital anomalies are 42% higher than naturally conceived children (Williams et al. 2010). Some of the birth defect increases are likely a result of increased preterm birth and multiple-birth offspring, but some of the increased rates may be because the parents are less likely to terminate the pregnancy upon discovery of the birth defect (Williams et al. 2010). Offspring born through ART are 1.4 to 2 times more likely to have low birth weight (Green 2004). Multiple births, which occur in over half of pregnancies resulting from ART, are correlated with preterm offspring and low birth weight (Fountain and Krulewitch 2002). However, even singletons from ART tend to have low birth weight, which may be the result of the "vanishing twin" (Jauniaux et al. 2013). With IVF, which accounts for nearly three-fourths of all ART procedures, multiple fertilized ova are inserted into the female, and often these fertilized ova do not take and become the "vanishing twin" (Jauniaux et al. 2013).

Preterm offspring, regardless of the cause of the preterm birth, have higher rates of birth defects than full-term offspring. Honein et al. (2009), by examining data from thirteen states that were collected between 1995 and 2000, found that birth defects, such as reduced limbs, microcephaly, spina bifida, congenital hydrocephaly, and cleft palate, were twice as likely in infants born at between twenty-four to thirty-six weeks than in infants born between thirty-seven and forty-one weeks. In very early births (twenty-four to thirty-one weeks), the risk of birth defects was five times higher than in full-term births (Honein et al. 2009).

Multiple births, also regardless of cause of the multiple births, lead to pre-term births and an increase in birth defects (Reefhuis et al. 2009). Twins and multiple births, for instance, have a higher rate of neural tube defects (Jauni-aux et al. 2013) and cleft palates (Reefhuis et al. 2009).

Smoking and Alcohol

Teen mothers are more likely to have offspring with birth defects than non-teen mothers, but these stem from lifestyle choices, such as smoking, poor diet, and drug use (Wang et al. 2012a). When controlling for lifestyle and socioeconomics, research has found that birth defects are more common in older mothers than in younger mothers. According to the CDC (2013), most birth defects occur in the first trimester, which makes healthy lifestyle deci-sions essential even before one knows that one is pregnant.

Tobacco use, mostly in the form of cigarettes, has been linked to clubfoot, cleft palate, craniosynostosis, and spina bifida (Engesæter 2006; Hackshaw et al. 2011). In a meta-analysis, Hackshaw et al. found that maternal smoking increased limb malformations by 26%. Seven thousand chemicals related to smoking can cross the placental barrier (Hackshaw et al. 2011). In developed countries, smoking rates have decreased. In the United Kingdom, 16.3% of pregnant females smoked in 1987; ten years later that statistic was down to 11.8%. The CDC (2013) reported that in the United States in 1965, smoking rates were as high as 42%, but in 2011 smoking had gone down to 18.9%. And in Canada, a similar decrease has led to a current smoking rate of 17%. Ac-cording to the Cancer Council of Australia, female smoking was highest in 1976 when a third of Australian females smoked; in 2010, just under 10% of females smoke. Due to the decrease in smoking, related birth defects should decrease too. Yet most smokers smoke during pregnancy (Ebrahim et al. 2000), and young mothers are more likely to smoke. Also, even with the smok-ing rates dropping in the developed world, developing countries are seeing an increase in smoking according to the WHO (2013).

Alcohol use trends are difficult to assess. The increase in higher alcohol content beverages, the increase in the size of the beverages, and the frequency of consumption make it difficult to determine whether alcohol use has in-creased. It has been suggested that alcohol use has dropped in the United States (Keyes et al. 2011), but some researchers have found that black and Hispanic populations have not seen a drop in alcohol consumption (Caetano and Clark 1998). According to the NIH (2013), 20% to 30% of pregnant fe-males consume alcohol, and 12% have binged on alcohol. Alcohol has become more available and common in Africa and Asia (WHO 2013). The increase in alcohol in developing nations has led to an increase in fetal alcohol syndrome and other birth defects according to the WHO (2013). Alcohol, for example, is the number one cause of birth defects in South Africa (WHO 2013). Alcohol consumption during pregnancy can lead to **midfacial anomalies**, growth retardation, learning disabilities, microcephaly, and malocclusion (usually an underbite). Although fetal alcohol syndrome (the suite of traits associated with offspring from mothers who drank alcohol during pregnancy) was first described in 1973 by Jones and Smith, it is sometimes difficult to assess which birth defects are caused by alcohol and which are caused by other substances, since **polysubstance abuse** is common among mothers who drink (Wekselman et al. 1995). Feldman et al. (2012), who examined a sample of nearly one thousand mothers and their offspring through the California Teratogen Information Service and Clinical Research Program (1978–2005), found that alcohol use had a linear relationship with risk of birth defects and that there is perhaps no safe threshold, but they mentioned the difficulty in assessing alcohol consumption since mothers may under-report their drinking. Interestingly, animal research on mice has suggested that folate or folic acid consumption reduces the negative effects of alcohol (Serrano et al. 2010).

Obesity and Diabetes

Pre-pregnancy obesity is on the rise in the United States; from 2003 to 2009, pre-pregnancy obesity increased from 17.6% to 20.5% (Fisher et al. 2013). In some states, pre-pregnancy obesity is as high as 38% (Reece 2008). According to the Pregnancy Risk Assessment Monitoring System, which contains data on over one hundred thousand females, African-Americans and Native Americans had pre-pregnancy obesity rates of around 29%, whereas about 20% of European-American females had pre-pregnancy obesity (Fisher et al. 2013). Asians and Pacific Islanders had the lowest rates at around 7%. Fisher et al. also reported that the rate of pre-pregnancy obesity was highest in females over the age of thirty-five and lowest in teens. Pre-pregnancy obesity is not just an American problem; in Australia in 1980, pre-pregnancy obesity was only 7.1%, but in 2000 the rate increased to 18.4% (Oddy et al. 2009). And in the United Kingdom, a quarter of adult females were obese as of 2007 (Rankin et al. 2010).

Pre-pregnancy obesity has been linked to an increased risk of spina bifida, cleft palates, limb reductions, and multiple anomalies. For example, using a sample of seven hundred, Oddy (2009) found that neural tube defects, cleft palates, and limb reductions were twice as frequent in the offspring of obese females as compared to infants who had normal or overweight mothers. Rankin et al. (2010) examined a large sample of over thirty thousand from five maternity wards in northeast England and found that congenital anomalies, such as cleft lip and palate, were more common in infants from obese and underweight mothers. Obesity increases the risk of spina bifida; some research has suggested that the increased odds ratio may be as high as 3.5 (Reece 2008). In a meta-analysis, infants of obese mothers have been found to have nearly two times the risk of neural tube defects compared to infants from normal-weight mothers, and infants of severely obese mothers have a three times greater risk of neural tube defects (Rasmussen et al. 2008). Although most studies show that overweight females do not have an increased risk of having an infant with congenital defects, Reece (2008) reported that the risk for multiple anomalies (in single infants) increases with a BMI between 25 and 29.9, which is defined as overweight.

There are several hypotheses as to why obesity increases birth defect risk; one possibility includes the fact that glucose metabolism goes awry in obese individuals. Animal studies have found that hyperglycemia, which is excess glucose in the blood, is **teratogenic** (Rasmussen et al. 2008). In diabetics, glucose builds up in the blood, and cells do not get the glucose they need for energy and growth. Gestational diabetes and pre-gestational type II diabetes increase the risk of many types of birth defects, especially neural tube defects (Correa et al. 2008; Rasmussen et al. 2008; Reece 2008). Obesity is also highly correlated with type II diabetes. Obesity and diabetes cause an additive effect on the risk of having a child with a birth defect, such as spina bifida, congenital hydrocephaly, and anencephaly (Reece 2008). Obese mothers who have diabetes have a two to four times greater risk of getting a child with a birth defect than non-obese and non-diabetic mothers (Correa et al. 2012; Reece 2008). Some research has suggested that diabetes and not obesity is the real culprit for the increased risk of birth defects (see Reece 2008). One-third of diabetic mothers were undiagnosed at the time of their pregnancy (Correa et al. 2012). In addition, six out of ten pregnant females with diabetes got pregnant unplanned and have difficulty following a **glycemic**-controlled diet (Correa et al. 2012). Some animal studies have demonstrated that vitamin C, vitamin E, and vitamin B-9 (in either folate or folic acid) reduces birth defect risk in offspring of diabetic females (Correa et al. 2012; Wentzel 2009).

Although there is evidence that susceptibility to diabetes and obesity is related to genetic variance (Lupo et al. 2012), there is also evidence that lifestyle, especially diet, plays a role in obesity and type II diabetes (Lim and Cheng 2011; Parker et al. 2012). Diets with high glycemic intakes have been linked to increased birth defect risks when the mothers are obese (Parker et al. 2012). Two ways to reduce the risks of birth defects from obesity and diabetes include losing weight before pregnancy and decreasing weight gain during pregnancy (Lim and Cheng 2011).

Folic Acid Fortification and Neural Tube Defects
Another possible cause of the birth defect link with pre-pregnancy obesity may be that obese females require more vitamin B-9 than normal females. Folate

and folic acid prevents neural tube defects, and thus heavier females may need more than the recommended dose of four hundred micrograms that was determined for normal-weight females (Rasmussen et al. 2008; Reece 2008).

Neural tube defects, such as spina bifida, anencephaly, and **encephalocele**, can often be prevented through the use of vitamin B-9 (Wu et al. 2007). How folate and folic acid prevent neural tube defects is not completely understood (Blom et al. 2006). According to some researchers, 70% of neural tube defects can be prevented with folic acid fortification (Chen 2008). Folate is the most common deficient vitamin in females (Wu et al. 2007). As early as the 1960s, researchers noted that low levels of folate led to complications during pregnancy (Bar-Oz et al. 2008). Folate is needed during cell and tissue growth, and deficiency in the early stages of pregnancy seems to impair cell growth and replication (Bar-Oz et al. 2008). In the early 1990s, several landmark studies that were led by Sir Nicholas Wald, Smithells, Cziezel, and Dudas shaped folate consumption recommendations and the folic acid fortification of foods. These studies and others that used the Medical Research Council (MRC) Vitamin Study Group employed randomized double-blind placebo methods from thirty-three centers. The outcome of the studies was that women of childbearing ages should consume four hundred micrograms of folate or folic acid a day (Bar-Oz et al. 2008; Wu et al. 2007). Best practice involves vitamin B-9 (in either folate or folic acid forms) consumption at least one month prior to conception and throughout the first trimester. This is because neural tubal development occurs during the first three to four weeks of pregnancy, which is often before females even realize they are pregnant (Wu et al. 2007).

Although folate is found in many foods, such as broccoli, lentils, beans, fruits, peas, nuts, and leafy green vegetables, it is difficult to obtain enough folate from diet. Thus, folic acid fortification was recommended (Bar-Oz et al. 2008; Wu et al. 2007). By 1998, folic acid fortification of wheat and cereals started in the United States; Canada soon followed with its fortification plan (Bar-Oz et al. 2008). Canada followed in part because food trade with the United States relied on it. Required folic acid fortification has been adopted by sixty-seven countries (Rick et al. 2012). Sixty-one countries fortify wheat flour, one country fortifies maize, and six countries fortify both wheat and maize (Rick et al. 2012). The March of Dimes (2013) has proposed that the United States should start fortifying corn flour since Hispanics, who have higher rates of neural tube defects, often consume corn flour in their traditional foods. In some countries, folic acid fortification resulted in drastic reductions of neural tube defects; for example, Argentina saw a 45% reduction, Canada's reduction was 46%, and Saudi Arabia's neural tube defect rate dropped by 60% (Rick et al. 2012). In China, the four hundred micrograms a day of folic acid led to an 85% reduction in spina bifida in the highest-risk areas (Oakley 2009). However, most countries do not enforce the fortification practices (Oakley 2009), and in some places, such as Peru, the fortification is too low to reduce birth defects (Rick et al. 2012). Even with the widespread use of folic acid, no country can yet claim that they have eliminated folate deficiency spina bifida or anencephaly (Oakley 2009). Furthermore, only about 10% of spina bifida is actually prevented worldwide (Oakley 2009). Some of this is a result of countries that do not have folic acid fortification; Hungary, for example, does not have folic acid fortification laws, but they have seen a drop in neural tube defects in live births as a result of screening (which may increase the number of cases discovered) and termination of pregnancies (Szabó et al. 2013). Even in places where folic acid fortification is required, neural tube defects have not reduced as drastically as one may have hoped.

Despite the evidence that vitamin B-9 reduces neural tube defects, vitamin B-9 consumption is still not universal for women capable of getting pregnant. In Texas, Canfield et al. (2006) found that 28% of the women they interviewed knew that folic acid prevented birth defects. Also, Yang et al. (2012b) examined a sample of 16,541 households from the Shaanxi Province in China and found that only 11% of pregnant females took folic acid supplements or a multivitamin. In Honduras, Wu et al. (2007) found that only about 45% of females knew about the health benefits of folic acid for their offspring, and among health care workers, only about half of them advised females of childbearing age to take folic acid. These statistics are some of the reasons that neural tube defects caused by folate deficiency remain a significant cause of infant mortality and morbidity. But there are other reasons for vitamin B-9 deficiency. In the United States, the low-carb diet fad of the late 1990s and early 2000s led to a decrease in foods that supplied folic acid. Additionally, the increase in BMI may result in an epidemic of females not getting enough folic acid.

Beyond folate consumption patterns, there may be other reasons for the incomplete amelioration of neural tube defects. In the Netherlands, after education about folate started in 1995, folate consumption increased from 1995 to 1998 compared to 1988 (van der Pal-de Bruin et al. 2003). By 1996, one in five pregnant females used folic acid supplements, and by 1998 folic acid consumption increased by 35%. However, the reduction in birth defects was not as significant as hoped. Perhaps genetics prevented the reduction; single nucleotide polymorphisms have been implicated in the prevention of folate metabolism (Alfarra et al. 2011; Chen 2008; Mitchell et al. 2004). The genetic causes of folate deficiency are still being researched (Alfarra et al. 2011; Blom et al. 2006).

CONCLUSIONS

Skeletal records and clinical databases on birth defects do not share many similarities; spina bifida and cleft palate information are two exceptions. In both the skeletal record and the clinical records, these are the most common birth defects. However, bioarchaeological remains of spina bifida are usually limited to spina bifida occulta, which is unlikely to be considered a defect in clinical databases. Major birth defects, such as Down syndrome and spina bifida aperta or cystica, are rarely reported in paleopathology articles. The lack of data from the past reflects the rarity of any one birth defect and the unlikely preservation of skeletal remains from miscarriages, stillbirths, or infant deaths. Thus the understanding of past congenital defects is limited, but the understanding of current defects is aided by large databases. Even though most anomalies have an unknown etiology, one can discuss factors that may lead to increasing healthy births. In future skeletal records, osteologists may be even less likely to see birth defects due to pregnancy termination from screening, but the defects that do occur in the developed world are likely to be a result of obesity, diabetes, and older mothers. Some of the adverse effects can be reduced by taking prenatal vitamins with folic acid, which older mothers and women who plan pregnancy do take. There is still room for improvement; women of childbearing age should consume sufficient vitamin B-9 even if they are not planning on having a child, just in case. In the developing nations, birth defects caused by inbreeding seem to be dropping as a result of female education and urbanization; conversely, increases in tobacco and alcohol consumption may erase any gains made by increased outbreeding.

CHAPTER 10

The Next Fifty Years?

THROUGHOUT THIS BOOK, YOU MAY HAVE NOTICED THAT WHEN ONE PROBlem is solved, another problem emerges or an old problem arises from the solution. For example, drugs to treat osteoporosis may lead to a new type of femoral shaft fracture. Or avoiding teen pregnancy has driven motherhood to the other end of the spectrum, and now birth defects associated with older mothers are a growing problem. And the rise and education of the middle class has led to well-intentioned but misinformed food fads, such as feeding infants soy milk, which can lead to protein deficiency.

You may have also noticed many modern problems differ between the developing nations and the developed nations. Interestingly, however, some of the bony outcomes are the same. For instance, whereas excess food has led to the obesity epidemic in North America, food shortages have led to malnutrition in Africa. In both cases, the result is short stature. For North America it is a result of early puberty, and for Africa it is a result of stunted growth. Yet other obesity-related bone disorders, such as the increase in knee osteoarthritis, are not seen nearly as much in the developing nations. The developing world, too, continues to cope with issues that the developed world no longer has problems with, such as parasite infections as a result of over-population, unsanitary living conditions, and weakened immune systems. In the developed world, some diseases are nevertheless making a comeback due to the anti-vaccination movement and diet fads. Plus, new problems have arisen due to use of medicines, such as tooth loss as a result of dry mouth. And then again, fortified foods and prenatal screenings have reduced some of the problems related to birth defects and maybe have even led to a healthier next generation. Furthermore, many of the developed nations' problems are a result of the aging population; osteoarthritis and osteoporosis, for example, increase in older adults, but one can hardly suggest that this newfound longevity is a bad thing.

Yet trends covered in chapters 1 through 9 may not continue into the future; the world is a dynamic place, and new concerns may arise while old problems may be solved. In the next few pages, I will summarize the trends that I predict will occur over the next five decades. The trends include (1) increasingly older populations; (2) increased understanding of genetic causes of diseases and increased medical interventions; (3) leveling off of obesity rates in the West, but an increase in obesity rates in other parts of the world; (4) urbanization of the developing nations, in part due to a rise in technology; (5) a continuation of fortification of foods, especially in Asia and South America, even though small pockets of Western people will dislike this movement; (6) increasingly older parents; and (7) a return of past diseases.

OLDER POPULATIONS

Throughout the previous chapters, many bone disorders were indicators of childhood stresses, but as childhood becomes safer through vaccinations, fortified foods, and child labor laws, one may expect a drop in childhood stress indicators, such as Harris lines, enamel hypoplasia, and short stature. With increasing longevity, osteologists may expect a greater interest in understanding how aging affects bone biology. The US senior population is increasing at a faster rate than other age groups. Longevity is also increasing globally. According to the WHO in 2011, life expectancy globally was seventy years of age, and although life expectancy ranged from forty-six years of age in Sierra Leone males to eighty-six years of age in Japanese females, many of the lowered life expectancies in developing nations were the result of childhood mortality. When examining years of life remaining at age sixty, in the developed countries the average was twenty-four years, which put the life expectancy up to eighty-four years old, whereas in developing countries at age sixty, one could expect to live an extra seventeen years, which put the life expectancy to seventy-seven years of age.

As populations age and there are more individuals living into their eighties and nineties, doctors can expect an increase in osteoarthritis and osteoporosis, although the individuals who experience osteoarthritis may not be the same people who experience osteoporosis, if the theory that osteoarthritis is inversely related to bone loss is correct. Older populations may not be equal in health; for instance, in North America, a lifelong struggle with obesity may leave the elderly with greater osteoarthritis disability, more diabetes-related ulcers (and amputations), and an overall result of a more dependent old-age population; whereas fit older individuals—perhaps especially from East Asia because of their healthier diets and more active lifestyles—are more likely to continue an active lifestyle. Regardless of the population differences, increased longevity will likely drive the Western world to conduct research on the genetics of aging, especially to answer questions concerning why some individuals have healthier bones and joints than others, and perhaps individualized medicines that use one's own cells to fight diseases to help those with osteoarthritis pain, diabetic ulcers, and osteoporotic fractures.

GENETIC AND MEDICAL ADVANCES

Many bone disorders are inherited. For instance, twin studies have revealed that genetics can explain half or more of the variance in bone mineral density in the lower back, Schmorl's nodes, osteoarthritis, and even caries risk. I suspect that in the future, osteologists will learn about the influence of genetics on many traits. Furthermore, perhaps information about genetics and environmental interactions will be revealed. These advances may allow doctors to cater their treatments to individuals.

Before individualized medicine will become a reality, I suspect that the number of pharmaceuticals will continue to increase. The increased medication of the developed nations' populations may result in an increase in tooth loss due to dry mouth. It may also cause an increase in other problems, such as fractured femoral shafts as a result of osteoporosis medications. But perhaps one of the biggest problems that will arise in the next fifty years as a result of medical intervention is the increase in unnecessary surgeries to treat pains, such as back pains. Evidence from past populations has illustrated that some bony pathologies are actually just variations; every osteophyte, stress fracture, or hernia does not need to be treated. Spondylolysis, for instance, occurs in as much as 50% of some bioarchaeological populations; thus it may just be a harmless and painless way to increase flexibility. Some vertebral pathologies, such as Schmorl's nodes, may be painful in the short term but may not result in long-term pain. Measures to treat pain will likely increase as more doctors put the blame on inheritance rather than lifestyle, when it may be that osteoarthritis, Schmorl's nodes, and other bony features are inherited, but that pain attributed to these traits is not inherited. Excess weight, for example, has been linked to increased pain sensitivity. Obesity is perhaps the biggest issue in health in the next five decades.

OBESITY

Obesity has been mentioned multiple times in the preceding chapters. For example, osteoporosis is less likely in obese individuals, but a whole slew of other deleterious traits arise in obese individuals. Obese children reach puberty earlier and cease growth earlier, which results in shorter adults. Additionally, obesity and a lack of activity may result in an increase in Blount's disease in children. In adults, obese individuals have an increased risk of periodontal disease, back pain, osteoarthritis pain, and diabetes. Diabetes can lead to foot ulcers, which in turn can lead to amputations. Obese mothers-to-be have an increased risk of having a child with birth defects, such as spina bifida and cleft palates. In the developed countries, obesity in the next fifty years will likely level off because of awareness of the problems that have been fueled by over-sized portions, inactivity, and unhealthy diets. However, the obesity epidemic is not over, and in developing nations there will likely be an increase in obesity over the next several decades. Places such as Egypt, Morocco, Uzbekistan, Algeria, and Zambia, which have an increasing number of obese children, will face problems of earlier puberty that may even result in more teen mothers. Without fortified foods in these nations, obese mothers are even more likely to have offspring with birth defects. Furthermore, obesity in the developing nations will be driven by an increase in urbanization that results in a less active lifestyle and a greater reliance on high-calorie, low-nutrition foods.

URBANIZATION
More people are living in cities than ever before. The modern trend of urban-ization is in part a result of the rise in technology; in developing countries this has increased crowded and unsanitary conditions coupled with an increase in sedentary lifestyles. Early urbanization led to dismal living conditions in Eu-rope in the Industrial Age; crowded and unsanitary living conditions coupled with excessive workloads and a lack of sunlight (due in part to time spent in factories and in part to pollution) allowed contagious diseases, such as tuber-culosis and cholera, to flourish. The evidence of these conditions can be seen in European skeletal remains from the Industrial Revolution in the form of collapsed vertebrae and cranial lesions from tuberculosis, rickets from vitamin D deficiency, and general bone infections. Now, urban areas in the developed countries have better-educated and healthier populations on average than their rural neighbors. However, in the developing world, urban populations face a greater risk of diseases due to the same problems Europeans had last millennia. With over-crowded, unsanitary living conditions in Africa, Central and South America, and Asia, contagious diseases are making a comeback; doctors can expect a rise in tuberculosis, staph infections, and parasite loads. Anemia may increase as well due to parasites.

Nevertheless, it may be that urban locations—where increased wealth is not dependent on landownership or tribal connections—may lead to a new middle class that will demand changes in the local infrastructure. Many of these people will send their children to be educated, inbreeding (which has been found to result in supernumerary teeth, polydactyly, and spina bifida) will decrease, females will have their offspring at more opportune times, and they will likely have fewer children. Thus, population growth may slow down. Furthermore, these populations may eventually be successful in gaining ac-cess to clean water sources that are fluoridated and food that is fortified. Thus, healthier teeth that last a lifetime may become the norm. Since tooth loss and periodontal disease are linked to cardiovascular diseases, an increase in dental health will have wide-reaching effects. In the long run, perhaps urbanization will be a positive trend.

FOOD FORTIFICATION

In the next fifty years, I think that food fortification and modification will undergo major changes. In developed nations, perhaps folic acid fortification will increase both in quantity of fortification per food item and in what foods are fortified. The introduction of folic acid in 1998 reduced birth defects, especially neural tube defects like spina bifida, in the United States; this fortification of wheat in the US was followed by fortification of grains in sixty-seven countries. In the future, I think that the United States will likely add fortification to corn flour due to the increasing Hispanic population, which will hopefully bring the spina bifida rates of Hispanics down in line with their Caucasian counterparts. Plus, fortification may be increased to accommodate the increase in body size, which may then further reduce birth defects. In 1990, the average US body weight was about twenty pounds lighter than in 2002; this increase in body weight continued in the following decade (Mendes 2011). One may assume that people are consuming more food, and therefore consuming more folic acid naturally. Following this reasoning, additional fortification is not required, but the increase in calorie consumption has not been equal across all foods. Sugar and fat consumption has increased, but fortified foods, such as grains, milk, and orange juice, may actually be consumed less now. Fortification thus may need to be added to more foods and in greater quantities to the already fortified foods.

Developing nations have yet to use fortification effectively. With greater urbanization and fewer people relying on their own grown foods to feed themselves, government regulation of foods may increase. Fortification of rice in Asia, of corn in the Americas, and of a variety of grains in Africa will likely reduce birth defects. However, to be effective, fortification needs to be consistent and sufficient. In some places where fortification standards were too low, birth defect rates did not change as much as hoped.

Beyond fortification of foods with folic acid, genetically modified organisms (GMOs) were been introduced into human food systems as early as 1990. In the next fifty years, malnutrition in developing nations may increase, especially if urbanization spoils rare agricultural and pastoral land in countries like China, which arguably is not even a developing nation. Some of the problems arise from a lack of food, but more importantly there is a lack of nutrient-rich food, such as meats and dairy. Zinc and iron deficiencies, for example, are common and likely will continue to be common, especially if population growth does not cease. In the skeleton, the evidence of malnutrition is seen in stunted growth, and anemia specifically shows up as cribra orbitalia and porotic hyperostosis. Cribra orbitalia and porotic hyperostosis increased in many locations with the adoption of plant-based diets through agriculture, perhaps as a result of low-iron diets. Anemia's effects go beyond bones; cribra orbitalia and porotic hyperostosis are just indicators that these individuals would have been weak, sickly, and perhaps cognitively impaired.

One way to ameliorate the problems of malnutrition is to use GMOs. Genetically modified potatoes and rice have been produced to increase the nutrients of these staples (Wu et al. 2014). Yet, due to the perceived harmful effects of GMOs on individuals (as a result of rodent studies) and the environment, many nations have put a ban on GMOs. For example, Zaire's government refused genetically modified maize even when the people were faced with starvation from drought and crop failure (Wu et al. 2014). Whether the resistance to GMOs will increase or decrease is anyone's guess; in the United States, genetically modified foods have not been banned and are likely to increase in the future, although there has been a growing backlash.

INCREASING PARENTAL AGE

As education expectations increase and individuals take longer to start their financially independent lives, age at becoming a parent increases too. In the developed countries over the last thirty years, age at first birth has increased; the fertility myth, which is that becoming pregnant is incredibly easy and needs to be avoided when one is young and that one is never too old to become a parent, has led many to put off starting their families. I suspect that this trend will increase and spread globally as more females, especially in developing countries, attain an education. Increasing parental age has some benefits; for instance, older parents tend to be more conscientious about their health and lifestyle while planning their families. Alcohol consumption and cigarette smoking, for example, both have negative effects on developing fetuses, such as an increased chance of clubfoot or cleft palates, and older parents-to-be are more likely to abstain from or at least cut down on these vices. Furthermore, in the developed nations especially, older mothers-to-be are more likely to consume nutritious foods and take prenatal vitamins, which will decrease neural tube defects. The downside is that older mothers-to-be are more likely to be overweight, which can lead to a variety of birth defects. And of course chromosomal abnormalities are more likely to arise in mothers-to-be past the age of thirty-five years than in younger mothers. Even older fathers-to-be are linked to increased risks of cleft palates and limb deformities.

The increase in birth defect rates, however, does not necessarily translate to higher rates of individuals born with birth defects. In the West, increased screening has led to an increase in terminated pregnancies, and thus no increase in birth defect rates. Genetic testing has become less expensive, less risky, and more accurate. Multiple birth defects can now be detected with DNA tests rather than amniocentesis (Hayden 2014). Yet some people may be reluctant to terminate pregnancies, such as those with religious convictions against abortion. Furthermore, individuals who engage in ART seem to be less likely to terminate pregnancies. Using ART increases multiple births and birth defects. Multiple births and premature babies (which can be singletons or multiple births) also tend to have greater risk of spina bifida, cleft palate, and hydrocephaly. Couples who have tried to start their families unsuccessfully naturally for years may be hesitant to terminate a successful ART session. Furthermore, the perception that medical intervention can cure or treat all types of problems can lead to a false sense of security. Medical intervention can aid in helping premature or underweight babies survive, but their long-term outlook may include lifelong struggles (e.g., Irving et al. 2000). Finding a balance between successful careers, financial security, and starting a family is difficult and will likely become more difficult in the next fifty years. In the developed countries, the age of mothers-to-be is likely not going to get older, but in the developing countries these issues are just starting, and gains from fortification of food and a decrease of inbreeding may be erased with birth defects resulting from increased maternal and paternal age.

RETURN OF PAST DISEASES

In the next fifty years, what is old may come around once again. Diseases that were thought to be conquered through vaccination, such as measles and whooping cough, have resurged in the West. Childhood diseases can lead to lifelong deformities, such as atrophied limbs from poliomyelitis. Furthermore, childhood diseases can result in much energy spent on fighting infections rather than spent on growth. Harris lines and short stature can be the result, and these correspond to early cognitive impairment and later heart problems. The autism-vaccine link, which has now been fully discredited, is still being presented in a balanced manner by the media and online (Dixon and Clarke 2013). Furthermore, as more people act as their own researchers using the Internet, not being able to distinguish between valid sources and biased sources will reinforce people's own views. That is, people will search specifically for support for their view, which is referred to as confirmation bias, and on the Internet all views can be found, even many that are invalid.

Additionally, food fads, such as gluten-free diets, veganism, and low-carb diets, may result in an increase in birth defects due to low consumption of fortified foods. And these special diets can also lead to malnutrition. Medical doctors may see more children who are consuming soy milk and therefore growing up protein and vitamin D deficient. Doctors may need to be re-taught to recognize anemia, protein deficiency, and rickets as a result of the educated middle class using random websites to choose healthy diets rather than looking to more established and credentialed sites, such as the CDC and the NIH. Plus, as adults, more females may be increasing their early-onset osteoporosis risk by engaging in veganism, which can lead to Colles' fractures and vertebral collapse. Thus, early-onset osteoporosis may be an issue as it was in medieval Europe. This can lead to less able older females. However, these deficiencies can be cured with changes in diet, genetically modified foods, and supplements.

In the next fifty years, food transportation across borders will likely increase; recent research has shown that some rice from China is tainted with heavy metals, such as lead and mercury (Fang et al. 2014). Furthermore, spices from Mexico have been found to have greater contamination with *Salmonella* than spices from Europe, the United States, and Canada (van Doren et al. 2013). Doctors can thus expect more food-borne illnesses in the future. Furthermore, consumer preferences against pasteurization and preservatives will likely increase food contamination and thus infections from bacteria, such as *H. pylori*. These food-linked diseases in children will cause stunted growth and in adults may increase anemia. Maybe the pendulum will swing the other way, and a return to safe and fortified foods will become popular in both the developing and developed world.

CONCLUSIONS

The study of the past can help us understand current health problems, and this in turn will help in better future planning. Throughout this book, I have attempted to make relevant links between the past and the present. The links, however, are not only temporal—hopefully links between the different disciplines have also been established. Information from bioarchaeology, medical case studies, autopsy collections, animal research, evolutionary anatomy, and large clinical data sets are combined here to explain bone health. Physical anthropology is not just scholarly; it adds value to real-world issues, and I hope this book will start many discussions among current and future health care providers, medical anthropologists, and the wider public. Understanding human evolution, biology, and lifestyle choices can help people to make wise health decisions. Furthermore, educated choices can reduce, though not eliminate, the risks of osteoarthritis, back pains, and osteoporosis. These bone disorders may seem minor, but understanding them can lead us to avoid unnecessary and unhelpful medical procedures and lead to a long and independent life with as little disability as possible.

Acronyms

aDNA	ancient deoxyribonucleic acid
ART	artificial reproductive technologies
BMD	bone mineral density
BMI	body mass index
CDC	Centers for Disease Control and Prevention
CHIRPP	Canadian Hospitals Injury Reporting and Prevention Program
DEXA	dual energy X-ray absorptiometry
DPA	dual photon absorptiometry
CT	computer tomography
DISH	diffuse idiopathic skeletal hyperostosis
GMO	genetically modified organism
HIV	human immunodeficiency disease
IVF-ICSI	in vitro fertilization–intracytoplasmic sperm injection
MRI	magnetic resonance imaging
MRSA	methicillin-resistant staphylococcus aureus
NHANES	National Health and Nutrition Examination Surveys
NIH	National Institutes of Health
NSAIDS	non-steroidal anti-inflammatory drugs
OA	osteoarthritis
OCD	osteochondritis dissecans
SNP	single nucleotide polymorphism
TB	tuberculosis
WHO	World Health Organization

Glossary

abrasion: Wear produced by interaction between teeth and other materials.

acetabular dysplasia: A shallow and abnormally directed hip socket.

acid: Substances whose aqueous solutions are characterized by a sour taste and have a pH of less than 7.

acupuncture: A Chinese medical procedure in which specific body areas are pierced with fine needles to relieve pain.

adipocyte: Fat cell.

allele: An alternative form of the same gene.

Allen's rule: A biological rule posited by Joel A. Allen in 1877 that states animals from colder climates usually have shorter distal elements in relation to proximal elements; animals from warmer climates have comparatively longer distal elements.

alveolar: Pertaining to the tooth socket.

amelia: Congenital absence of a limb; other forms of *-melia* include *hemimelia*, which refers to the absence of one part of a limb, such as the fibula, but not the tibia.

amenorrhea: Absence of menstruation.

amino acid: Biologically important organic compound that produces proteins.

androgenic: Related to male steroid hormones.

anemia: Decrease in the number of red blood cells as a result of iron deficiency or blood loss.

anencephaly: The congenital absence of a major portion of the brain, skull, and scalp.

anisotropic: The property of being directionally dependent; isotropy is the opposite and implies identical properties in all direction.

ankylosing spondylitis: Fusion of the spine as a result of chronic inflammatory disease.

anorexia nervosa: Eating disorder characterized by severe restriction and irrational fear of weight gain.

antemortem: Occurring before death; occurring after death is postmortem.

anterolisthesis: Vertebral body slippage when the body flips forward; the opposite is retrolisthesis when the body slips back.

anthropometric: The measurement of humans.

antibiotic: Medicine that is aimed at treatment of bacterial infections.

aperture: An anatomical hole or opening.

arthritis: Inflammation of a joint that usually causes pain, swelling, and stiffness; it can be a result of infection, trauma (also known as secondary arthritis), degenerative changes (also known as osteoarthritis), autoimmune disorders (also known as rheumatoid arthritis), or other causes.

articular surface: Any surface of a skeletal formation that makes normal direct contact with another skeletal structure as part of a synovial joint.

asymmetry: Dissimilarity in corresponding parts; symmetry is the opposite term.

asymptomatic: Experiencing no symptoms.

atresia: Congenital absence or closure of a normal body opening or tubular structure.

atrophy: Wasting away.

attrition: Tooth loss caused by abrasive foods or grinding of teeth.

autosomal recessive: A pattern of inheritance in which both copies of an autosomal (non-sex chromosome) gene must be abnormal for a genetic condition or disease to occur. When both parents have one abnormal copy of the same gene, they have a 25% chance with each pregnancy that their offspring will have the disorder.

avulsion: The forcible separation of a body part from the main structure.

bacterium: Single-celled prokaryotic microorganism; plural: bacteria.

betel: An East Indian plant that is chewed as a stimulant.

bilateral: Pertaining to both sides; opposite is unilateral.

bipedalism: Moving around on two limbs.

bisphosphonates: Calcium-regulating drugs that inhibit bone resorption.

bone resorption: Loss of bone resulting from osteoclasts.

calcaneus: The heel bone; plural: calcanei.

calcification: Hardening of bone with calcium.

calcium: The chemical element used in mineralization of bone, teeth, and shells.

calculus: Also known as tartar; calcium phosphate and carbonate, with organic matter, deposited on tooth surfaces; plural: calculus.

canaliculus: Small tube or channel; plural: canaliculi.

carbohydrate: A compound, such as cellulose, sugar, or starch, that contains only carbon, hydrogen, and oxygen, and is a major part of diets to produce energy.

cardiovascular disease: Disease that affects the heart and blood vessels.

caries: Also known as cavities; tooth decay as a result of bacteria that feed on carbohydrates on the tooth surface; plural: caries.

caries sicca: Cranial bone decay as a result of syphilis.

cariogenic: Food that increases the risk of dental caries.

catch-up growth: An acceleration of the growth rate following a period of growth retardation caused by a secondary deficiency.

celiac disease: A disease of the digestive system that damages the small intestine and interferes with the absorption of nutrients from food.

cementum: Bonelike connective tissue covering the root of a tooth and assisting in tooth support.

-cephaly: A suffix meaning head, cranium, skull, or brain.

cerebral palsy: A group of congenital non-progressive disorders of movement and posture caused by abnormal development of, or damage to, motor control centers of the brain.

cessation: The stopping of an event or thing.

chlorpyrifos: A toxic, suspected endocrine disruptor that is used as an insecticide.

chondrification centers: A site of cartilage formation in the body.

chondrocyte: Cartilage cell.

chondromalacia: Malformation of cartilage.

chronic: Persisting for a long time.

cleft palate: Congenital fissure of median line of palate.

cloaca: An opening in a diseased bone.

clubfoot: A congenital deformity of the foot that causes it to twist into an abnormal position.

collagen: Fibrous protein that supports the connective tissues.

Colles' fracture: A fracture of the forearm that is a result of catching one's fall.

condyle: A rounded articular surface or projection of bone.

congenital: Existing at or before birth.

consanguineous: Related by blood.

coronary heart disease: Chronic cardiac disabilities resulting from insufficient supply of oxygenated blood to the heart.

cortical bone: Compact bone of limb shafts that surround the medullary/marrow cavity.

craniosynostosis: Premature closure of the cranial sutures.

-cyte: Suffix for a cell.

Dark Ages: Also known as the medieval period; the period from about the late fifth century AD to about AD 1000.

deep vein thrombosis: A condition that results in a blood clot in a major vein that usually develops in the legs and/or pelvis; it can be fatal.

demineralization: Excessive removal of minerals from bone and tooth tissues.

dentin: The main hard tissue of the tooth.

deworming: To cure of parasitic worms.

diaphysis: The shaft of a long bone; plural: diaphyses.

dizygotic twins: Non-identical twins; two individuals born at the same time from the same parents, but as a result of two ova being fertilized by two separate sperm cells; monozygotic twins are identical twins that are conceived from one ovum and one sperm cell.

double-blind study: An experiment in which information about the experiment that may lead to bias is concealed from the tester and the subject.

Down syndrome: The most common cause of mental retardation and malformation; it occurs because of the presence of an extra twenty-first chromosome and is also known as trisomy 21. Down syndrome was first described in 1866 by Dr. John L. H. Down.

dwarfism: A condition of abnormally short stature. An achondroplastic dwarf is one with a large head and short limbs; chondroplastic dwarfs are proportioned similarly to non-dwarfs.

dysplasia: Any abnormality of development.

E. coli: Abbreviation for the *Escherichia coli* bacteria, which is normally found in the human gastrointestinal tract and exists as numerous strains, some of the strains causing diarrhea.

eburnation: Degeneration of bone into a hard ivory-like shiny mass, usually a result of arthritis.

eczema: An inflammatory condition of the skin that can be caused by autoimmune diseases.

edentulous: Without teeth.

enamel: Hard, thin, translucent substance covering and protecting the dentin of a tooth crown and made of calcium salts.

enamel hypoplasia: A defect in which the enamel of the teeth is deficient as a result of stress.

encephalocele: A congenital gap in the skull that often causes a protrusion or hernia of brain material.

endochondral: Formed or occurring within cartilage.

endosteum: The tissue lining the medullary cavity of a bone.

environmental: Pertaining to causes that are not genetic.

epidemic: An outbreak of a disease in a particular region.

epiphyseal plate: Also called a growth plate, it is made up of a thin layer of cartilage between the epiphysis, a secondary bone-forming center, and the bone shaft. Epiphyseal plates remain open until late adolescence.

epiphysis: The end of a long bone; plural: epiphyses.

erosion: Wear of dental tissue by chemicals, such as citric acid.

etiology: Cause or origin of diseases.

external auditory meatus: Ear canal.

fluoride: A mineral that is utilized in maintaining healthy teeth and bones.

folate: Vitamin B-9 that is essential in preventing neural tube defects; folic acid is man-made vitamin B-9 that is used to fortify foods.

fungus: Eukaryotic organism that reproduces by spores; plural: fungi.

gestation: Period of fetal development from conception to birth; also known as pregnancy.

gingivitis: Inflammation of the gums (i.e., the soft tissue surrounding the teeth).

glenohumeral: The shoulder joint.

glucosamine: An amino acid sugar found in chitin, cell membranes, and mucopolysaccharides, which are used to treat osteoarthritis.

glycemic: Pertaining to the level of glucose (i.e., sugar) in the blood.

greenstick fracture: A fracture found in non-brittle bones in which one side of a bone is broken and the other side is bent.

Harris lines: Lines most visibly seen on tibial X-rays that are related to growth interruptions.

helminth: A parasitic worm.

hemophilia: Congenital disease that results in a lack of blood clotting and excessive bleeding.

hemorrhaging: To bleed profusely.

hernia: The protrusion of an organ or other bodily structure through the wall that normally contains it.

hetero-: A prefix meaning different or varied; opposite is homo.

heterodonts: Teeth that are different in shape and function; opposite is homodonts.

heterotopic: Displacement of an organ or other body part to an abnormal location.

histological analysis: Examination at the microscopic level.

homeostasis: Stability in the normal physiological states of the organism.

hominins: Family of primates that includes past and present humans.

hormones: Chemicals produced by glands in the body that circulate in the blood and control the actions of cells and organs.

hydrocephaly: An abnormal expansion of cavities within the brain caused by the accumulation of cerebrospinal fluid.

hypo-: A prefix meaning deficient or lacking; opposite is hyper.

in utero: Inside the womb; pertaining to or occurring before birth.

Industrial Age: A period of history characterized by the replacement of hand tools with power-driven machines that began in the mid-1700s in England and later in other countries; the Industrial Age was replaced by the Information Age in the late twentieth century.

insulin: A hormone that helps regulate metabolism of carbohydrates and fats.

interobserver error rates: Data variation between different researchers taking the same data or measurement; intraobserver error rates are variations when a single researcher collects the same data multiple times.

-itis: Used as a suffix to mean inflammation.

kyphosis: Extreme curvature of the upper back; lordosis is extreme curvature of the lower back.

lactation: Yielding milk from the mammary glands to enable breastfeeding.

lacuna: A small cavity in the bone matrix; plural: lacunae.

lamella: A thin plate of bone tissue; plural: lamellae.

Last Glacial Maximum: Refers to a period between 26,500 and 19,000–20,000 years ago in the Earth's climate history when ice sheets were most extensive and temperatures were lowest, which marked the peak of the last glacial period.

leptin: A protein that affects feeding and hunger.

Levant: The countries bordering on the eastern Mediterranean Sea from Turkey to Egypt.

ligament: A band of tough fibrous tissues connecting bones or cartilage at a joint or muscle or supporting an organ.

lipping: Bony overgrowths that form a ring around the joint in arthritis.

longitudinal studies: Studies that follow subjects for a long period of time.

Looser's zone: An area of transverse lines seen on X-rays that are similar to Harris lines.

lumbarization: Sacral development of the fifth lumbar vertebra; when lumbar vertebrae are fused to the sacrum, then it is called sacralization.

lymph cells: Cells that function in the development of immunity and include two specific types, B cells and T cells.

-lysis: A suffix that refers to separation.

mal-: Prefix meaning bad or incorrect.

malocclusion: Faulty contact between teeth when the mouth is closed.

mandible: Lower jawbone.

maxilla: Upper jawbone.

medullary cavity: The inner core of long bones.

menarche: The start of the menstrual cycle; a time when females first become fertile.

meniscus: A disk of cartilage that acts as a cushion between the ends of bones in a joint.

metabolism: The sum of chemical changes involved in the function of nutrition; basically, the way the body utilizes food to maintain life.

micro-: Prefix meaning small; opposite is *macro-*.

Middle Ages: The period in European history from AD 476 to 1453.

midfacial anomalies: Physical deformities that occur in the nasal and cheek areas.

minerals: Inorganic elements, such as calcium, iron, potassium, sodium, or zinc, that are essential to nutrition.

molars: Commonly referred to as cheek teeth; these are teeth toward the back of the mouth.

multicusps: A tooth with multiple raised surfaces.

multifactorial: Involving, dependent on, or controlled by several factors.

mycobacterium: A genus of bacteria that causes contagious diseases; the most common form is tuberculosis (abbreviated as TB); a relatively rare form is leprosy.

nasomaxillary: The region of the face that involves the nose and upper jaw; also sometimes called the rhinomaxillary.

necrosis: Death of body tissue that happens when there is not enough blood flowing to the tissue.

Neolithic Revolution: The changes that occurred prior to and with the onset of agriculture; around ten thousand years ago.

neoteny: Retention of juvenile characteristics in the adults of a species.

neural tube defects: Birth defects of the brain, spine, or spinal cord.

neuropathy: Conditions that occur when nerves that carry messages to the brain and spinal cord from the rest of the body are damaged or diseased.

New World: North America, Central America, and South America.

occlusal surface: Biting surface of a tooth.

Old World: Europe, Asia, and Africa.

omnivorous: Consuming a wide variety of foods.

oral contraceptives: Pills to prevent pregnancy; birth control pills.

Osgood-Schlatter: A painful swelling of the bump on the upper part of the shinbone, just below the knee.

osteitis: Any type of bone inflammation.

osteo-: Prefix meaning pertaining to bone.

osteoblasts: Cells responsible for the synthesis and mineralization of bone during both bone formation and remodeling.

osteoclasts: Cells responsible for the breakdown or absorption of bone tissue back into the body.

osteocytes: A general term for bone cells.

osteogenesis: The formation of bone.

osteogenesis imperfecta: A congenital disease that causes extremely fragile bones; also known as brittle bone disease.

osteoids: The unmineralized and organic part of bone that forms prior to the calcification of bone tissue.

osteology: The study of bones.

osteomalacia: The softening of the bones in adults due to vitamin D deficiency.

osteomyelitis: A bone infection that has penetrated the medullary cavity.

osteopenia: A bone condition characterized by bone loss that is less severe than osteoporosis.

osteophytes: Bone spurs that usually develop around joints and the edges of bone.

osteoporosis: A bone condition characterized by bone loss severe enough to increase risk of fractures.

pamidronates: A group of drugs that are injected to prevent bone loss.

pandemic: An epidemic occurring on a scale that crosses international boundaries, usually affecting a large number of people.

panmixia: Random mating.

parasites: An organism that lives on or in a host and gets its food from or at the expense of its host.

parity: The fact of having given birth.

pathogens: Organisms that cause diseases.

pathological: Relating to pathology or disease.

peak bone mass: The greatest amount of bone mass in an individual's life.

perio-: Prefix pertaining to tissue surrounding or close to the teeth.

periodontal: Related to tissues surrounding the teeth.

periodontal disease: Disease of the gum tissue.

periodontitis: Inflammation of the tissues surrounding the teeth.

periosteum: A membrane that covers the outer surface of all bones except at the joints.

periostitis: Inflammation of the periosteum.

phocomelia: Congenital defect wherein the bones of one or more limbs are missing or shortened.

phosphorus: A mineral/chemical element that is essential in the formation of bones and teeth.

placebo: A substance containing no medication and prescribed or given to reinforce a patient's expectation to get well.

plasticity: The ability to change and adapt.

pleiotropy: In genetics, when one gene codes for many traits.

poly-: Prefix meaning many.

polydactyly: Having more than the normal numbers of fingers or toes.

polygenic: In genetics, when many genes code for one trait.

polysubstance abuse: Overuse or misuse of multiple drugs, alcohol, and tobacco.

porosity: The state of being porous.

porotic hyperostosis: The expansion of the diploë and thinned outer table of bones of the cranial vault (i.e., the parietals), usually associated with iron deficiency.

post-: Prefix meaning after; opposite is *pre-*; *pre-* can also refer to early.

post-menopausal: The period of time for a female who has experienced a year without menstruation.

postmortem: Occurring after death.

Pott's disease: A condition that causes arthritis-like deformation of the spine as a result of tuberculosis.

primary dentition: The first set of teeth that develop in the human oral cavity; also referred to as milk teeth and deciduous teeth.

primates: The order of mammals that contains prosimians, monkeys, apes, and humans.

proliferative: Growing or increasing in number rapidly.

proteins: Complex molecules made of amino acids; also one of the three nutrients used as energy sources by the body. Proteins are essential components of the muscle, skin, and bones.

protozoan: Single-celled organisms, some types of which can cause disease; plural: protozoa.

pseudofracture: A condition in which an X-ray shows formation of new bone with thickening of periosteum at the site of an injury to bone.

psychoses: Mental disorders.

puberty: Period during which children undergo the process of sexual maturation.

pulp: Soft or fleshy tissue; in teeth, the inner substance of the tooth that contains the arteries, veins, lymphatic tissue, and nerves.

pus: Thick, opaque, and usually yellowish-white fluid that is formed as a result of an infection; it is composed of white blood cells, tissue debris, and microorganisms.

pyogenic: Producing pus.

quadrupedalism: Moving habitually on four limbs.

respiratory disease: Any disease that affects the lungs, nose, mouth, pharynx, larynx, trachea, the chest wall, diaphragm, or the neuromuscular system that provides the power for breathing.

rickets: A softening-of-bone disease that occurs in children as a result of vitamin D deficiency.

robusticity: Strength of skeletal elements.

sagittal clefting: On the vertebrae, abnormal fissure or opening that is the result of failure of parts to fuse during embryonic development.

Salmonella: The name of a group of bacteria, which is one of the most common causes of food poisoning in the United States.

Schistosoma: A type of parasitic worm, also known as blood flukes. The illness they cause is also called schistosoma or bilharzia.

Schmorl's nodes: In the vertebrae, herniations of the intervertebral discs through the cartilaginous end plate and into the cancellous bone.

scoliosis: Abnormal spinal curvature of more than ten degrees.

scurvy: A disease that causes anemia as a result of vitamin C deficiency.

secondary dentition: The second set of teeth; also known as permanent teeth.

sedentary: Tending to sit or not engage in much physical activity.

sickle-cell anemia: A congenital form of iron deficiency as a result of abnormally shaped red blood cells; it is passed down through autosomal recessive inheritance and is known as one of the ethnic diseases.

smallpox: A contagious and sometimes fatal infectious disease caused by a virus. The only prevention is vaccination. The disease symptoms include raised bumps that appear on the face and body of an infected person.

socioeconomic status: The class or standing of an individual or group based on a variety of factors that include income, education, occupation, and social status.

spicules: Small or slender formation of bone, usually around a joint as a result of arthritis.

spina bifida: A birth defect in which the vertebrae do not form properly around the spinal cord.

spondylolisthesis: Displacement of a vertebra; slipped disk.

spondylolysis: Fracture of the vertebra in which the vertebral body is detached from the vertebral arch.

Staphylococcus: A bacterium that is frequently found in the human respiratory tract and on the skin, which can lead to infections; also known as staph.

stature: The height of someone.

stenosis: Abnormal narrowing in a tubular structure.

strain: An injury that involves a pulled muscle. Or, in bone remodeling, observed deformation when force (i.e., stress) is applied to bone.

Streptococcal: A bacteria that often results in a sore throat; also known as strep.

stress: Applied force to a bone.

stunting: A reduced growth rate in children.

subadult: An individual that has passed through the juvenile period but not yet attained typical adult characteristics.

subperiosteal: Tissue beneath the periosteum.

supernumerary teeth: A condition where more than the normal number of teeth are present.

supra-: Anatomical prefix meaning above.

surgical debridement: The process of removing dead tissue or foreign material from and around a wound to expose healthy tissue.

symmetric: In terms of pathology, occurring on both left and right sides.

syndesmophyte: A bony growth attached to a ligament.

syndrome: A set of symptoms occurring together.

tendon: A tough band of fibrous connective tissue that connects muscle to bone and is capable of withstanding tension.

teratogen: Any substance, organism, or process that causes malformations in a fetus.

thyroid: A butterfly-shaped gland that sits low on the front of the neck; it secretes hormones that influence metabolism, growth, development, and body temperature. During infancy and childhood, adequate thyroid hormones are crucial for brain development.

tibial tuberosity: An elevated prominence on the front of the shinbone that is the site for the distal patellar ligament.

trabecular bone: Bone in which the spicules form a latticework, with interstices filled with embryonic connective tissue or bone marrow; also known as cancellous or spongy bone. Found mainly on the end of long bones and the vertebral bodies.

trans-generational: In genetics, inheritance across multiple generations.

traumatic: A wound caused by an injury.

treponemal: Any of the syphilis or related diseases that are caused by the bacteria *Treponema sp*. It can be venereal (passed through sexual contact or body fluids), but there are non-venereal forms, such as yaws, endemic syphilis, and bejel.

type II diabetes: A form of diabetes mellitus that develops as a result of lifestyle, especially in adults and in obese individuals, that is characterized by hyperglycemia resulting from impaired insulin utilization coupled with the body's inability to compensate with increased insulin production. Also called non-insulin-dependent diabetes or adult-onset diabetes.

ulcer: A lesion that is eroding away the skin or mucous membrane.

urbanization: The act of urbanizing or taking on the characteristics of a city, usually the result of an increase in population density.

vascularize: To supply tissue with blood vessels.

vegan: An individual who does not eat any animal products, including meat, fish, dairy, or honey.

vegetarian: An individual who does not eat any meat.

vertebral body: The main part of the vertebrae that is placed anterior.

vertebral canal: The canal in successive vertebrae through which the spinal cord passes.

virus: An ultramicroscopic, metabolically inert, infectious agent that replicates only within the cells of living hosts; viruses are composed of an RNA or DNA core, a protein coat, and, in more complex types, a surrounding envelope.

vitamins: Any of various organic substances essential for normal growth and activity of the body.

Wolff's law: The principle that every change in the form or function of a bone leads to changes in its internal architecture and in its external form.

zebra lines: Multiple transverse sclerotic lines; often seen in X-rays of children who receive medicines to prevent bone loss.

Works Cited

Ackerman KE, Nazem T, Chapko D, Russell M, Mendes N, Taylor AP, Bouxsein ML, Misra M. 2011. Bone microarchitecture is impaired in adolescent amenorrheic athletes compared with eumenorrheic athletes and nonathletic controls. J Clin Endocrinol Metab 96:3123–3133.

Adair LS. 1999. Filipino children exhibit catch-up growth from age 2 to 12 years. J Nutr 129:1140–1148.

Addy M. 2008. Oral hygiene products: potential for harm to oral and systemic health? Periodontology 2000 48:54–65.

Adegboye AR, Twetman S, Christensen LB, Heitmann BL. 2012. Intake of dairy calcium and tooth loss among adult Danish men and women. Nutrition 28:779–784.

Adirim TA, Cheng TL. 2003. Overview of injuries in the young athlete. Sports Med 33:75–81.

Aebi M. 2012. Transition anomalies at the lumbosacral junctions. Eur Spine J 21:1223–1224.

Agadi JB, Madni NA, Nanjappa V, Govindaiah HK. 2010. Cryptococcal osteomyelitis of the skull in a patient with transient lymphopenia. Neurol India 58:300–302.

Agarwal SC, Dumitriu M, Tomlinson GA, Grynpas MD. 2004. Medieval trabecular bone architecture: the influence of age, sex, and lifestyle. Am J Phys Anthropol 124:33–44.

Ahmed ML, Ong KK, Dunger DB. 2009. Childhood obesity and the timing of puberty. Trends Endocrinol Metab 20:237–242.

Akpata ES. 1975. Molar tooth attrition in a selected group of Nigerians. Community Dent Oral Epidemiol 3:132–135.

Aksglaede L, Juul A, Olsen LW, Sørensen TI. 2009. Age at puberty and the emerging obesity epidemic. PLoS One 4:e8450.

Alfarra HY, Alfarra SR, Sadiq MF. 2011. Neural tube defects between folate metabolism and genetics. Indian J Hum Genet 17:126–131.

Alfonso-Durruty MP. 2011. Experimental assessment of nutrition and bone growth's velocity effects on Harris lines formation. Am J Phys Anthropol 145:169–180.

Aligne CA, Moss ME, Auinger P, Weitzman M. 2003. Association of pediatric dental caries with passive smoking. JAMA 289:1258–1264.

Alirol E, Getaz L, Stoll B, Chappuis F, Loutan L. 2011. Urbanisation and infectious diseases in a globalised world. Lancet Infect Dis 11:131–141.

Allali F, Maaroufi H, Aichaoui SE, Khazani H, Saoud B, Benyahya B, Abouqal R, Hajjaj-Hassouni N. 2007. Influence of parity on bone mineral density and peripheral fracture risk in Moroccan postmenopausal women. Maturitas 57:392–398.

Allen KD, Oddone EZ, Coffman CJ, Keefe FJ, Lindquist JH, Bosworth HB. 2010. Racial differences in osteoarthritis pain and function: potential explanatory factors. Osteoarthritis Cartilage 18:160–167.

Aloisi AM, Albonetti ME, Carli G. 1994. Sex differences in the behavioural response to persistent pain in rats. Neurosci Lett 179:79–82.

Álvarez-Díaz P, Alentorn-Geli E, Steinbacher G, Rius M, Pellisé F, Cugat R. 2011. Conservative treatment of lumbar spondylolysis in young soccer players. Knee Surg Sports Traumatol Arthrosc 19:2111–2114.

Ameen S, Stau L, Ulrich S, Vock P, Ballmer F, Anderson SE. 2005. Harris lines of the tibia across centuries: a comparison of two populations, medieval and contemporary in Central Europe. Skeletal Radiol 34:279–284.

Anastasiou E, Mitchell PD. 2013. Paleopathology and genes: investigating the genetics of infectious diseases in excavated human skeletal remains and mummies from past populations. Gene 528:33–40.

Andersen JG, Manchester K, Roberts C. 1994. Septic bone changes in leprosy: a clinical, radiological and palaeopathological review. Int J Osteoarchaeol 4:21–30.

Anderson JJ, Felson DT. 1988. Factors associated with osteoarthritis of the knee in the first National Health and Nutrition Examination Survey (HANES I): evidence for an association with overweight, race, and physical demands of work. Am J Epidemiol 128:179–189.

Anderson T. 2003. A medieval example of sagittal cleft or "butterfly" vertebra. Int J Osteoarchaeol 13:352–357.

Anderson T, Carter AR. 1995. An unusual osteitic reaction in a young medieval child. Int J Osteoarchaeol 5:192–195.

Ang DC, Ibrahim SA, Burant CJ, Kwoh CK. 2003. Is there a difference in the perception of symptoms between African Americans and whites with osteoarthritis? J Rheumatol 30:1305–1310.

Angel JL. 1966. Early skeletons from Tranquility, California. Smithson Contr Anthropol 2:1–19.

Anson C, Rothschild B, Naples V. 2012. Soft tissue contributions to pseudopathology of ribs. Adv Anthropol 2:57–63.

Antón SC, Polidoro GM. 2000. Prehistoric radio-ulnar synostosis: implications for function. Int J Osteoarchaeol 10:189–197.

Aragón-Sánchez J, Lázaro-Martínez JL, Hernández-Herrero C, Campillo-Vilorio N, Quintana-Marrero Y, García-Morales E, Hernández-Herrero MJ. 2012. Does osteomyelitis in the feet of patients with diabetes really recur after surgical treatment? Natural history of a surgical series. Diabet Med 29:813–818.

Aragón-Sánchez J, Lázaro-Martínez JL, Quintana-Marrero Y, Hernández-Herrero MJ, García-Morales E, Cabrera-Galván JJ, Beneit-Montesinos JV. 2009. Are diabetic foot ulcers complicated by MRSA osteomyelitis associated with worse prognosis? Outcomes of a surgical series. Diabet Med 26:552–555.

Arcini C, Frölund P. 1996. Two dwarves from Sweden: a unique case. Int J Osteoarchaeol 6:155–166.

Armfield JM, Spencer AJ, Roberts-Thomson KF, Plastow K. 2013. Water fluoridation and the association of sugar-sweetened beverage consumption and dental caries in Australian children. Am J Public Health 103:494–500.

Arnold WH, Naumova EA, Koloda VV, Gaengler P. 2007. Tooth wear in two ancient populations of the Khazar Kaganat Region in the Ukraine. Int J Osteoarchaeol 17:52–62.

Arora M, Weuve J, Weisskopf MG, Sparrow D, Nie H, Garcia RI, Hu H. 2009. Cumulative lead exposure and tooth loss in men: the Normative Aging Study. Environ Health Perspect 117:1531.

Arriaza BT. 1997. Spondylolysis in prehistoric human remains from Guam and its possible etiology. Am J Phys Anthropol 104:393–397.

Arriaza BT, Merbs CF, Rothschild BM. 1993. Diffuse idiopathic skeletal hyperostosis in Meroitic Nubians from Semna South, Sudan. Am J Phys Anthropol 92:243–248.

Arthritis Foundation. http://www.arthritis.org.

Avlund K, Schultz-Larsen K, Christiansen N, Holm-Pedersen P. 2011. Number of teeth and fatigue in older adults. J Am Geriatr Soc 59:1459–1464.

Babul S, Nolan S, Nolan M, Rajabali F. 2007. An analysis of sport-related injuries: British Columbia Children's Hospital Emergency Department 1999–2003. Int J Inj Contr Saf Promot 14:192–195.

Bach G, Shilling A. 2008. Research update: combating the overuse epidemic. Parks & Recreation 43:24–27.

Baird PA, Sadovnick AD, Yee IML. 1991. Maternal age and birth defects: a population study. Lancet 337:527–530.

Bajwa NS, Toy JO, Ahn NU. 2012. L5 pedicle length is increased in subjects with spondylolysis: an anatomic study of 1072 cadavers. Clin Orthop Relat Res 470:3202–3206.

Bajwa NS, Toy JO, Young EY, Cooperman DR, Ahn NU. 2013. Disk degeneration in lumbar spine precedes osteoarthritic changes in hip. Am J Orthop 42:309–312.

Balarajan R, McDowall M. 1985. Mortality from congenital malformations by mother's country of birth. J Epidemiol Community Health 39:102–106.

Balassy C, Hörmann M. 2008. Role of MRI in paediatric musculoskeletal conditions. Eur J Radiol 68:245–258.

Barnes GP, Parker WA, Lyon Jr TC, Drum MA, Coleman GC. 1992. Ethnicity, location, age, and fluoridation factors in baby bottle tooth decay and caries prevalence of Head Start children. Public Health Rep 107:167–173.

Barnes WS, Wilkinson D, Lalumandier JA. 2002. Comparing caries in two similar populations seventy years apart. J Dent Res 81:A92–A92.

Bar-Oz B, Koren G, Nguyen P, Kapur BM. 2008. Folate fortification and supplementation—are we there yet? Reprod Toxicol 25:408–412.

Basha B, Rao DS, Han ZH, Parfitt AM. 2000. Osteomalacia due to vitamin D depletion: a neglected consequence of intestinal malabsorption. Am J Med 108:296–300.

Bass WM, Gregg JB, Provost PE. 1974. Ankylosing spondylitis (Marie Strumpel disease) in historic and prehistoric Northern Plains Indians. Plains Anthropol 19:303–305.

Baum T, Joseph GB, Nardo L, Virayavanich W, Arulanandan A, Alizai H, Carballido-Gamio J, Nevitt MC, Lynch J, McCulloch CE, Link TM. 2013. MRI-based knee cartilage T2 measurements and focal knee lesions correlate with BMI—36 month follow-up data from the Osteoarthritis Initiative. Arthritis Care Res 65:23–33.

Beauchesne P, Agarwal SC. 2011. Age-related cortical bone maintenance and loss in an Imperial Roman population. Int J Osteoarchaeol. doi:10.1002/oa.1303.

Beltrán-Aguilar ED, Barker LK, Canto MT, Dye BA, Gooch BF, Griffin SO, Jaramillo F, Kingman A, Nowjack-Raymer R, Selwitz RH, Wu T. 2005. Surveillance for dental caries, dental sealants, tooth retention, edentulism, and enamel fluorosis—United States, 1988–1994 and 1999–2002. MMWR Surveill Summ 54:1–43.

Benazzi S, Nguyen HN, Schulz D, Grosse IR, Gruppioni G, Hublin JJ, Kullmer O. 2013. The evolutionary paradox of tooth wear: simply destruction or inevitable adaptation? PLoS One 8:e62263.

Bennell KL, Malcolm SA, Khan KM, Thomas SA, Reid SJ, Brukner PD, Ebeling PR, Wark JD. 1997. Bone mass and bone turnover in power athletes, endurance athletes, and controls: a 12-month longitudinal study. Bone 20:477–484.

Bennett KA. 1972. Lumbo-sacral malformations and spina bifida occulta in a group of protohistoric Modoc Indians. Am J Phys Anthropol 36:435–440.

Berbesque JC, Marlowe FW, Pawn I, Thompson P, Johnson G, Mabulla A. 2012. Sex differences in Hadza dental wear patterns. Hum Nat 23:270–282.

Bergink AP, van Meurs JB, Loughin J, Arp PP, Fang Y, Hofman A, van Leeuwen J, van Duijn CM, Uitterlinden AG, Pol HAP. 2003. Estrogen receptor α gene haplotype is associated with radiographic osteoarthritis of the knee in elderly men and women. Arthritis Rheum 48:1913–1922.

Bernard T, Wilde F, Aluoch M, Leaverton PE. 2010. Job-related OA of the knee, foot, hand, and cervical spine. J Occup Environ Med 52:33–38.

Biau DJ, Leclerc P, Marmor S, Zeller V, Graff W, Lhotellier L, Leonard P, Mamoudy P. 2012. Monitoring the one year postoperative infection rate after primary total hip replacement. Int Orthop 36:1155–1161.

Bittles AH, Black ML. 2010. Consanguineous marriage and human evolution. Annu Rev Anthropol 39:193–207.

Blanco RA, Acheson RM, Canosa C, Salomon JB. 1974. Height, weight, and lines of arrested growth in young Guatemalan children. Am J Phys Anthropol 40:39–47.

Blau S. 2001. Limited yet informative: pathological alterations observed on human skeletal remains from third and second millennia BC collective burials in the United Arab Emirates. Int J Osteoarchaeol 11:173–205.

Blau S. 2005. An unusual aperture in a child's calvaria from western Central Asia: differential diagnoses. Int J Osteoarchaeol 15:291–297.

Blau S, Yagodin V. 2005. Osteoarchaeological evidence for leprosy from western Central Asia. Am J Phys Anthropol 126:150–158.

Bliddal H, Christensen R. 2006. The management of osteoarthritis in the obese patient: practical considerations and guidelines for therapy. Obesity Rev 7:323–331.

Blom HJ, Shaw GM, den Heijer M, Finnell RH. 2006. Neural tube defects and folate: case far from closed. Nat Rev Neurosci 7:724–731.

Blondiaux G, Blondiaux J, Secousse F, Cotten A, Danze PM, Flipo RM. 2002. Rickets and child abuse: the case of a two year old girl from the 4th century in Lisieux (Normandy). Int J Osteoarchaeol 12:209–215.

Boel LWT, Ortner DJ. 2013. Skeletal manifestations of skin ulcer in the lower leg. Int J Osteoarchaeol 23:303–309.

Bogin B. 1999. Evolutionary perspective on human growth. Annu Rev Anthropol 29:109–153.

Bogin B, Silva MIV, Rios L. 2007. Life history trade-offs in human growth: adaptation or pathology? Am J Hum Biol 19:631–642.

Boldsen JL. 2001. Epidemiological approach to the paleopathological diagnosis of leprosy. Am J Phys Anthropol 115:380–387.

Boldsen JL. 2005. Analysis of dental attrition and mortality in the medieval village of Tirup, Denmark. Am J Phys Anthropol 126:169–176.

Bonmatí A, Gómez-Olivencia A, Arsuaga JL, Carretero JM, Gracia A, Martínez I, Lorenzo C, Bérmudez de Castro JM, Carbonell E. 2010. Middle Pleistocene lower back and pelvis from an aged human individual from the Sima de los Huesos site, Spain. Proc Natl Acad Sci USA 107:18386–18391.

Borse N, Sleet DA. 2009. CDC childhood injury report: patterns of unintentional injuries among 0–19 year olds in the United States, 2000–2006. Fam Community Health 32:189.

Brender JD, Felkner M, Suarez L, Canfield MA, Henry JP. 2010. Maternal pesticide exposure and neural tube defects in Mexican Americans. Ann Epidemiol 20:16–22.

Brickley M, Mays S, Ives R. 2005. Skeletal manifestations of vitamin D deficiency osteomalacia in documented historical collections. Int J Osteoarchaeol 15:389–403.

Brickley M, Mays S, Ives R. 2007. An investigation of skeletal indicators of vitamin D deficiency in adults: effective markers for interpreting past living conditions and pollution levels in 18th and 19th century Birmingham, England. Am J Phys Anthropol 132:67–79.

Bridges PS. 1989. Changes in activities with the shift to agriculture in the southeastern United States. Curr Anthropol 30:385–394.

Bridges PS. 1992. Prehistoric arthritis in the Americas. Annu Rev Anthropol 21:67–91.

Bridges PS. 1994. Vertebral arthritis and physical activities in the prehistoric southeastern United States. Am J Phys Anthropol 93:83–93.

Bromiker R, Glam-Baruch M, Gofin R, Hammerman C, Amitai Y. 2004. Association of parental consanguinity with congenital malformations among Arab newborns in Jerusalem. Clin Genet 66:63–66.

Brooks BK, Southam SL, Mlady GW, Logan J, Rosett M. 2010. Lumbar spine spondylolysis in the adult population: using computed tomography to evaluate the possibility of adult onset lumbar spondylosis as a cause of back pain. Skeletal Radiol 39:669–673.

Brooks ST, Hohenthal WD. 1963. Archaeological defective palate crania from California. Am J Phys Anthropol 21:25–32.

Broulik PD, Vondrova J, Ruzicka P, Sedlacek R, Zima T. 2010. The effect of chronic alcohol administration on bone mineral content and bone strength in male rats. Physiol Res 59:599–604.

Brunader R, Shelton DK. 2002. Radiologic bone assessment in the evaluation of osteoporosis. Am Fam Physician 65:1357–1365.

Bucci C, Gallotta S, Morra I, Fortunato A, Ciacci C, Iovino P. 2013. Anisakis, just think about it in an emergency! Int J Infect Dis 17:e1071–e1072.

Buckler JMH, Green M. 2008. The growth of twins between the ages of 2 and 9 years. Ann Hum Biol 35:75–92.

Buckley HR, Dias GJ. 2002. The distribution of skeletal lesions in treponemal disease: is the lymphatic system responsible? Int J Osteoarchaeol 12:178–188.

Burger H, van-Daele PL, Odding E, Valenburg HA, Hofman A, Grobbee DE, Schütte HE, Birkenhäger JC, Pols HAP. 1996. Association of radiographically evident OA with higher bone mineral density and increased bone loss with age, the Rotterdam Study. Arthritis Rheum 39:81–86.

Burke KL. 2012. Schmorl's nodes in an American military population: frequency, formation, and etiology. J Forensic Sci 57:571–577.

Burnwal R, Neogi DS, Ortho SSD. 2012. Tubercular osteomyelitis of distal ulna presenting as epiphyseal injury. Mædica 7:247–250.

Burr DB. 1997. Muscle strength, bone mass, and age-related bone loss. J Bone Miner Res 12:1547–1551.

Burt BA, Kolker JL, Sandretto AM, Yuan Y, Sohn W, Ismail AI. 2006. Dietary patterns related to caries in a low-income adult population. Caries Res 40:473–480.

Caballero B. 2002. Global patterns of child health: the role of nutrition. Ann Nutr Metab 46:3–7.

Caetano R, Clark CL. 1998. Trends in alcohol consumption patterns among whites, blacks and Hispanics: 1984 and 1995. J Stud Alcohol Drugs 59:659–668.

Caine D, DiFiori J, Maffulli N. 2006. Physeal injuries in children's and youth sports: reasons for concern? Br J Sports Med 40:749–760.

Callewaert F, Sinnesael M, Gielen E, Boonen S, Vanderschueren D. 2010. Skeletal sexual dimorphism: relative contribution of sex steroids, GH-IGF1, and mechanical loading. J Endocrino 207:127–135.

Campbell JR, Estey MP. 2013. Metal release from hip prostheses: cobalt and chromium toxicity and the role of the clinical laboratory. Clin Chem Lab Med 51:213–220.

Canale ST. 2007. Osteochondrosis or epiphysitis and other miscellaneous affections. In: Canale ST, Beatty JH, editors. Campbell's Operative Orthopaedics. 11th ed. Philadelphia (PA): Mosby Elsevier. p. 1133–1143.

Canfield MA, Przybyla SM, Case AP, Ramadhani T, Suarez L, Dyer J. 2006. Folic acid awareness and supplementation among Texas women of childbearing age. Prev Med 43:27–30.

Cardoso HFV, Garcia S. 2009. The not-so-Dark Ages: ecology for human growth in medieval and early twentieth century Portugal as inferred from skeletal growth profiles. Am J Phys Anthropol 138:136–147.

Cardoso HFV, Heuzé Y, Júlio P. 2010. Secular change in the timing of dental root maturation in Portuguese boys and girls. Am J Hum Biol 22:791–800.

Cardoso HFV, Padez C. 2008. Changes in height, weight, BMI and in the prevalence of obesity among 9- to 11-year-old affluent Portuguese schoolboys, between 1960 and 2000. Ann Hum Biol 35:624–638.

Carel JC, Lahlou N, Roger M, Chaussain JL. 2004. Precocious puberty and statural growth. Hum Reprod Update 10:135–147.

Carvalho NF, Kenney RD, Carrington PH, Hall DE. 2001. Severe nutritional deficiencies in toddlers resulting from health food milk alternatives. Pediatrics 107:e46–e46.

Case DT, Hill RJ, Merbs CF, Fong M. 2006. Polydactyly in the prehistoric American Southwest. Int J Osteoarchaeol 16:221–235.

Caspari R, Lee S. 2004. Older age becomes common late in human evolution. Proc Natl Acad Sci USA 101:10895–10900.

Cederroth CR, Auger J, Zimmermann C, Eustache F, Nef S. 2010. Soy, phyto-oestrogens and male reproductive function: a review. Int J Androl 33:304–316.

Cei S, D'Aiuto F, Duranti E, Taddei S, Gabriele M, Ghiadoni L, Graziani F. 2012. Third molar surgical removal: a possible model of human systemic inflammation: a preliminary investigation. Eur J Inflamm 10:149–152.

Centers for Disease Control and Prevention [CDC], Injury Prevention & Control. http://www.cdc.gov.

Cettour-Rose P, Samec S, Russell AP, Summermatter S, Mainieri D, Carrillo-Theander C, Montani JP, Seydoux J, Rohner-Jeanrenaud F, Dulloo AG. 2005. Redistribution of glucose from skeletal muscle to adipose tissue during catch-up fat: a link between catch-up growth and later metabolic syndrome. Diabetes 54:751–756.

Chalmers J, Conacher WDH, Gardner DL, Scott PJ. 1967. Osteomalacia—a common disease in elderly women. J Bone Joint Surg Br 49:403–423.

Chan CW, Peng P. 2011. Failed back surgery syndrome. Pain Med 12:577–606.

Chapple S. 2005. Sex inequality in the Māori population in the prehistoric, proto-historic and early historic eras in a Trans-Polynesian context. J Pac Hist 40:1–21.

Charadram N, Austin C, Trimby P, Simonian M, Swain MV, Hunter N. 2013. Structural analysis of reactionary dentin formed in response to polymicrobial invasion. J Struct Biol 181:207–222.

Chattah NLT, Smith P. 2006. Variation in occlusal wear of two Chalcolithic populations in the southern Levant. Am J Phys Anthropol 130:471–479.

Chen CP. 2008. Syndromes, disorders and maternal risk factors associated with neural tube defects (VI). Taiwan J Obstet Gynecol 47:267–275.

Chen MR, Huang JI, Victoroff BN, Cooperman DR. 2010. Fracture of the clavicle does not affect arthritis of the ipsilateral acromioclavicular joint compared with the contralateral side an osteological study. J Bone Joint Surg Br 92:164–168.

Chen X, Clark JJ. 2011. Multidimensional risk assessment for tooth loss in a geriatric population with diverse medical and dental backgrounds. J Am Geriatr Soc 59:1116–1122.

Chiu JF, Lan SJ, Yang CY, Wang PW, Yao WJ, Su IH, Hsieh CC. 1997. Long-term vegetarian diet and bone mineral density in postmenopausal Taiwanese women. Calcif Tissue Int 60:245–249.

Chow RH, Harrison JE, Notarius C. 1987. Effect of two randomised exercise programmes on bone mass of healthy postmenopausal women. BMJ 295:1441–1444.

Christian JC, Yu PL, Slemenda CW, Johnston Jr CC. 1989. Heritability of bone mass: a longitudinal study in aging male twins. Am J Hum Genet 44:429–433.

Chudá EP, Dörnhöferová M. 2011. Hyperostosis frontalis interna—a find in women individual from modern times (St. Martin Cathedral, Spisska Kapitula, Slovakia). Antrowebzin 2:97–101.

Cinar AB, Oktay I, Schou L. 2013. Relationship between oral health, diabetes management and sleep apnea. Clin Oral Investig 17:967–974.

Cleveland Museum of Natural History [CMNH], Hamann-Todd Osteological Collection. http://www.cmnh.org/site/ResearchandCollections/PhysicalAnthropology/Collections/Hamann-ToddCollection.aspx.

Clifton CE, Raffel S, Rytand D. Memorial Resolution Arthur William Meyer (1873–1966). Stanford Historical Society. http://histsoc.stanford.edu/pdfmem/MeyerA.pdf.

Cloke DJ, Khatri M, Pinder IM, McCaskie AW, Lingard EA. 2008. 284 press-fit Kinemax total knee arthroplasties followed for 10 years: poor survival of uncemented prostheses. Acta Orthop 79:28–33.

Clothier B, Stringer M, Jeffcoat MK. 2007. Periodontal disease and pregnancy outcomes: exposure, risk and intervention. Best Pract Res Clin Obstet Gynaecol 21:451–466.

Coates PS, Fernstrom JD, Fernstrom MH, Schauer PR, Greenspan SL. 2004. Gastric bypass surgery for morbid obesity leads to an increase in bone turnover and a decrease in bone mass. J Clin Endocrinol Metab 89:1061–1065.

Colella C. 2003. Understanding failed back surgery syndrome. Nurse Pract 28:31–43.

Colen S, Haverkamp D, Mulier M, van den Bekerom MP. 2012. Hyaluronic acid for the treatment of osteoarthritis in all joints except the knee. BioDrugs 26:101–112.

Collaer JW, McKeough DM, Boissonnault WG. 2006. Lumbar isthmic spondylolisthesis detection with palpation: interrater reliability and concurrent criterion-related validity. J Man Manip Ther 14:22–29.

Collier R. 2000. Regulation of rBST in the US. AgbioForum 3:156–163.

Collins VR, Muggli EE, Riley M, Palma S, Halliday JL. 2008. Is Down syndrome a disappearing birth defect? J Pediatr 152:20–24.

Colmenero JD, Ruiz-Mesa JD, Sanjuan-Jimenez R, Sobrino B, Morata P. 2013. Establishing the diagnosis of tuberculous vertebral osteomyelitis. Eur Spine J 22:S579–586.

Commandre FA, Taillan B, Gagenerie F, Zakarian H, Lescourgues M, Fourre JM. 1988. Spondylolysis and spondylolisthesis in young athletes: 28 cases. J Sports Med Phys Fitness 28:104–107.

Confavreux CB. 2011. Bone: from a reservoir of minerals to a regulator of energy metabolism. Kidney Int Suppl 79:S14–S19.

Cooper C, Campbell L, Byng P, Croft P, Coggon D. 1996. Occupational activity and the risk of hip osteoarthritis. Ann Rheum Dis 55:680–682.

Cope DJ, Dupras TL. 2011. Osteogenesis imperfecta in the archeological record: an example from the Dakhleh Oasis, Egypt. Int J Paleopath 1:188–199.

Cope JM, Berryman AC, Martin DL, Potts DD. 2005. Robusticity and osteoarthritis at the trapeziometacarpal joint in a Bronze Age population from Tell Abraq, United Arab Emirates. Am J Phys Anthropol 126:391–400.

Corbet EF, Leung WK. 2011. Epidemiology of periodontitis in the Asia and Oceania regions. Periodontology 2000 56:25–64.

Correa A, Gilboa SM, Besser LM, Botto LD, Moore CA, Hobbs CA, Cleves MA, Riehle-Colarusso TJ, Waller KD, Reece EA. 2008. Diabetes mellitus and birth defects. Am J Obstet Gynecol 199:237.e1–237.e9.

Correa A, Gilboa SM, Botto LD, Moore CA, Hobbs CA, Cleves MA, Riehle-Colarusso TJ, Waller KD, Reece EA. 2012. Lack of periconceptional vitamins or supplements that contain folic acid and diabetes mellitus–associated birth defects. Am J Obstet Gynecol 206:218.e1–218.e13.

Corruccini RS, Yap Potter RH, Dahlberg AA. 1983. Changing occlusal variation in Pima Amerinds. Am J Phys Anthropol 62:317–324.

Craig WJ. 2009. Health effects of vegan diets. Am J Clin Nutr 89:1627S–1633S.

Crosby J. 2009. Osteoarthritis: managing without surgery. J Fam Pract 58:354–361.

Cross NA, Hillman LS, Allen SH, Krause GF, Vieira NE. 1995. Calcium homeostasis and bone metabolism during pregnancy, lactation, and postweaning: a longitudinal study. Am J Clin Nutr 61:514–523.

Crowder C, Austin D. 2005. Age ranges of epiphyseal fusion in the distal tibia and fibula of contemporary males and females. J Forensic Sci 50:1001–1007.

Cucina A, Tiesler V. 2003. Dental caries and antemortem tooth loss in the Northern Peten area, Mexico: a biocultural perspective on social status differences among the Classic Maya. Am J Phys Anthropol 122:1–10.

Cunha BA. 2002. Osteomyelitis in elderly patients. Clinical Infectious Diseases 35:287–293.

Cunningham CA, Stephen A. 2010. The appearance of Harris lines at the iliac crest. Axis: The Online Journal of CAHId 2:13–21.

Curate F. 2008. A case of os odontoideum in the palaeopathological record. Int J Osteoarchaeol 18:100–105.

Curate F, Assis S, Lopes C, Silva AM. 2011. Hip fractures in the Portuguese archaeological record. Anthropol Sci 119:87–93.

Currey JD. 2012. The structure and mechanics of bone. J Mater Sci 47:41–54.

Dabernat H, Crubézy É. 2010. Multiple bone tuberculosis in a child from predynastic Upper Egypt (3200 BC). Int J Osteoarchaeol 20:719–730.

Dandona P, Okonofua F, Clements RV. 1985. Osteomalacia presenting as pathological fractures during pregnancy in Asian women of high socioeconomic class. BMJ 290(6471):837.

Daniel TM. 2000. The origins and precolonial epidemiology of tuberculosis in the Americas: can we figure them out? Unresolved issues. Int J Tuberc Lung Dis 4:395–400.

Dar G, Masharawi Y, Peleg S, Steinberg N, May H, Medlej B, Peded N, Hershkovitz I. 2010. Schmorl's nodes distribution in the human spine and its possible etiology. Eur Spine J 19:670–675.

Daragiu D-E, Ghergic DL. 2012. Correlation between malocclusion—oral habits—and socio-economic factors. Studia Universitatis 22:149–154.

D'Arcy Y. 2012. Pain and obesity. Nurs Manag 43:21–26.

Darton Y, Richard I, Truc MC. 2013. Osteomyelitis variolosa: a probable mediaeval case combined with unilateral sacroiliitis. Int J Paleopath. doi:10.1016/j.ijpp.2013.05.008.

Dawson H, Robson Brown K. 2012. Childhood tuberculosis: a probable case from late mediaeval Somerset, England. Int J Paleopath 2:31–35.

de la Cova C. 2011. Race, health, and disease in 19th-century-born males. Am J Phys Anthropol 144:526–537.

de Lucena GL, dos Santos Gomes C, Guerra RO. 2011. Prevalence and associated factors of Osgood-Schlatter syndrome in a population-based sample of Brazilian adolescents. Am J Sports Med 39:415–420.

de Onis M, Blössner M. 2000. Prevalence and trends of overweight among preschool children in developing countries. Am J Clin Nutr 72:1032–1039.

de Sá Pinto AL, De Barros Holanda PM, Radu AS, Villares SM, Lima FR. 2006. Musculoskeletal findings in obese children. J Paediatr Child Health 42:341–344.

Dean CL, Gabriel JP, Cassinelli EH, Bolesta MJ, Bohlman HH. 2009. Degenerative spondylolisthesis of the cervical spine: analysis of 58 patients treated with anterior cervical decompression and fusion. Spine J 9:439–446.

Den Hond E, Schoeters G. 2006. Endocrine disrupters and human puberty. Int J Androl 29:264–271.

Derevenski, JRS. 2000. Sex differences in activity-related osseous change in the spine and the gendered division of labor at Ensay and Wharram Percy, UK. Am J Phys Anthropol 111:333–352.

Deter CA. 2009. Gradients of occlusal wear in hunter-gatherers and agriculturalists. Am J Phys Anthropol 138:247–254.

Deyle GD, Denderson NE, Matekel RL, Ryder MG, Garber MB, Allison SC. 2000. Effectiveness of manual physical therapy and exercise in osteoarthritis of the knee. A randomized, controlled trail. Ann Intern Med 132:173–181.

Dhanoa A, Singh VA, Mansor A, Yusof MY, Lim KT, Thong KL. 2012. Acute haematogenous community-acquired methicillin-resistant *Staphylococcus aureus* osteomyelitis in an adult: case report and review of literature. BMC Infect Dis 12:270.

Dickel DN, Doran GH. 1989. Severe neural tube defect syndrome from the early Archaic of Florida. Am J Phys Anthropol 80:325–334.

DiGangi EA, Bethard JD, Sullivan LP. 2010. Differential diagnosis of cartilaginous dysplasia and probable Osgood-Schlatter's disease in a Mississippian individual from East Tennessee. Int J Osteoarchaeol 20:424–442.

Dillon CF, Hirsch R, Rasch EK, Gu Q. 2007. Symptomatic hand OA in the United States, prevalence and functional impairment estimates from the third U.S. National Health and Nutrition Examination Survey, 1991–1994. Am J Phys Med Rehabil 86:12–21.

d'Incau E, Couture C, Maureille B. 2012. Human tooth wear in the past and the present: Tribological mechanisms, scoring systems, dental and skeletal compensations. Arch Oral Biol 57:214–229.

Dixon G, Clarke C. 2013. The effect of falsely balanced reporting of the autism-vaccine controversy on vaccine safety perceptions and behavioral intentions. Health Educ Res 28:352–359.

Djurić-Srejić M, Roberts C. 2001. Paleopathological evidence of infectious disease in skeletal populations from later medieval Serbia. Int J Osteoarchaeol 11:311–320.

Dogra S, Narang T, Kumar B. 2013. Leprosy—evolution of the path to eradication. Indian J Med Res 137:15–35.

Domett K, Tayles N. 2006. Human biology from the Bronze Age to the Iron Age in the Mun River valley of northeast Thailand. In: Oxenham M, Tayles NG, editors. Bioarchaeology of Southeast Asia. Cambridge Studies in Biological and Evolutionary Anthropology 43. Cambridge: Cambridge University Press. p. 220–240.

Donnelly LF, Bisset III GS, Helms CA, Squire DL. 1999. Chronic avulsive injuries of childhood. Skeletal Radiol 28:138–144.

Dorsten LE, Hotchkiss L, King TM. 1996. Consanguineous marriage and early childhood mortality in an Amish settlement. Sociol Focus 29:179–185.

Dulloo AG. 2008. Thrifty energy metabolism in catch-up growth trajectories to insulin and leptin resistance. Best Pract Res Clin Endocrinol Metab 22:155–171.

Dumond H, Presle N, Terlain B, Mainard D, Loeuille D, Netter P, Pottie P. 2003. Evidence for a key role of leptin in osteoarthritis. Arthritis Rheum 48:3118–3129.

Duncan RC, Hay EM, Saklatvala J, Croft PR. 2006. Prevalence of radiographic osteoarthritis—it all depends on your point of view. Rheumatology 45:757–760.

Duncan WN, Stojanowski CM. 2008. A case of squamosal craniosynostosis from the 16th century southeastern United States. Int J Osteoarchaeol 18:407–420.

Dunn AJ, Campbell RS, Mayor PE, Rees D. 2008. Radiological findings and healing patterns of incomplete stress fractures of the pars interarticularis. Skeletal Radiol 37:443–450.

Dvornyk V, Liu XH, Shen H, Lei SF, Zhao LJ, Huang QR, Qin YJ, Jiang DK, Long JR, Zhang YY, Gong G, Recker RR, Deng HW. 2003. Differentiation of Caucasians and Chinese at bone mass candidate genes: implication for ethnic difference of bone mass. Ann Hum Genet 67:216–227.

Dwek JR. 2011. The radiographic approach to child abuse. Clin Orthop Relat Res 469:776–789.

Ebrahim SH, Floyd RL, Merritt II RK, Decoufle P, Holtzman D. 2000. Trends in pregnancy-related smoking rates in the United States, 1987–1996. JAMA 283:361–366.

Ecklund K, Jaramillo D. 2002. Patterns of premature physeal arrest MR imaging of 111 children. AJR Am J Roentgenol 178:967–972.

Edelson JG, Nathan H. 1986. Nerve root compression in spondylolysis and spondylolisthesis. J Bone Joint Surg Br 68:596–599.

Edmonds EW, Templeton KJ. 2013. Childhood obesity and musculoskeletal problems: editorial comment. Clin Orthop Relat Res 471:1191–1192.

Efe E, Sarvan S, Kukulu K. 2007. Self-reported knowledge and behaviors related to oral and dental health in Turkish children. Issues Compr Pediatr Nurs 30:133–146.

Egge MK, Berkowitz CD. 2010. Controversies in the evaluation of young children with fractures. Adv Pediatr 57:63–83.

Ehrlich GE. 2003. Back pain. J Rheumatol 67:26–31.

Eisenstein S. 1978. Spondylolysis. A skeletal investigation of two population groups. J Bone Joint Surg Br 60:488–494.

El-Din AM, El Banna RAE. 2006. Congenital anomalies of the vertebral column: a case study of ancient and modern Egypt. Int J Osteoarchaeol 16:200–207.

Engesæter LB. 2006. Increasing incidence of clubfoot: changes in the genes or the environment? Acta Orthop 77:837–838.

Erdal YS. 2006. A pre-Columbian case of congenital syphilis from Anatolia (Nicaea, 13th Century AD). Int J Osteoarchaeol 16:16–33.

Ericksen MF. 1982. Aging changes in thickness of the proximal femoral cortex. Am J Phys Anthropol 59:121–130.

Eriksson JG, Forsen T, Tuomilehto J, Winter PD, Osmond C, Barker DJ. 1999. Catch-up growth in childhood and death from coronary heart disease: longitudinal study. BMJ 318:427–431.

Erlandson MC, Sherar LB, Mirwald RL, Maffulli N, Baxter-Jones A. 2008. Growth and maturation of adolescent female gymnasts, swimmers, and tennis players. Med Sci Sports Exerc 40:34–42.

Ernst E. 2008. Chiropractic: a critical evaluation. J Pain Symptom Manage 35: 544–562.

Eshed V, Gopher A, Hershkovitz I. 2006. Tooth wear and dental pathology at the advent of agriculture: new evidence from the Levant. Am J Phys Anthropol 130:145–159.

Eshed V, Gopher A, Pinhasi R, Hershkovitz I. 2010. Paleopathology and the origin of agriculture in the Levant. Am J Phys Anthropol 143:121–133.

Etxebarria-Foronda I, Gorostiola-Vidaurrazaga L. 2013. Zebra lines: radiological repercussions of the action of bisphosphonates on the immature skeleton. Rev Osteoporos Metab Miner 5:39–41.

Eubanks JD, Cheruvu VK. 2009. Prevalence of sacral spina bifida occulta and its relationship to age, sex, race, and the sacral table angle: an anatomic, osteologic study of three thousand one hundred specimens. Spine 34:1539–1543.

Evensen JP, Øgaard B. 2007. Are malocclusions more prevalent and severe now? A comparative study of medieval skulls from Norway. Am J Orthod Dentofacial Orthop 131:710–716.

Évinger S, Bernert Z, Fóthi E, Wolff K, Kővári I, Marcsik A, Donoghue HD, O'Grady J, Kiss KK, Hajdu T. 2011. New skeletal tuberculosis cases in past populations from western Hungary (Transdanubia). HOMO 62:165–183.

Fang Y, Sun X, Yang W, Ma N, Xin Z, Fu J, Liu X, Liu M, Mariga AM, Zhu X, Hu Q. 2014. Concentrations and health risks of lead, cadmium, arsenic, and mercury in rice and edible mushrooms in China. Food Chem 147:147–151.

Farid R, Rezaieyazdi Z, Mirfeizi Z, Hatef MR, Mirheidari M, Mansouri H, Esmaelli H, Bentley G, Lu Y, Foo Y, Watson RR. 2010. Oral intake of purple passion fruit peel extract reduces pain and stiffness and improves physical function in adult patients with knee osteoarthritis. Nutr Res 30:601–606.

Farrer LA, Meaney FJ. 1985. An anthropometric assessment of Huntington's disease patients and families. Am J Phys Anthropol 67:185–194.

Fehily AM, Coles RJ, Evans WD, Elwood PC. 1992. Factors affecting bone density in young adults. Am J Clin Nutr 56:579–586.

Feldman HS, Jones KL, Lindsay S, Slymen D, Klonoff-Cohen H, Kao K, Rao S, Chambers C. 2012. Prenatal alcohol exposure patterns and alcohol-related birth defects and growth deficiencies: a prospective study. Alcohol Clin Exp Res 36:670–676.

Felson DT, Nevitt MC, Zhang Y, Aliabadi P, Baumer B, Gale D, Li W, Yu W, Xu L. 2002. High prevalence of lateral knee osteoarthritis in Beijing Chinese compared with Framingham Caucasian subjects. Arthritis Rheum 46:1217–1222.

Ferguson LA, Varnado JW. 2006. Syphilis: an old enemy still lurks. J Am Assoc Nurse Pract 18:49–55.

Fineberg HV, Hunter DJ. 2013. A global view of health—an unfolding series. N Engl J Med 368:78–79.

Fisher SC, Kim SY, Sharma AJ, Rochat R, Morrow B. 2013. Is obesity still increasing among pregnant women? Prepregnancy obesity trends in 20 States, 2003–2009. Prev Med 56:372–378.

Foldes AJ, Moscovici A, Popovtzer MM, Mogle P, Urman D, Zias J. 1995. Extreme osteoporosis in a sixth century skeleton from the Negev desert. Int J Osteoarchaeol 5:157–162.

Fontana L, Shew JL, Holloszy JO, Villareal DT. 2005. Low bone mass in subjects on a long-term raw vegetarian diet. Arch Intern Med 165:684–689.

Forand SP, Lewis-Michl EL, Gomez MI. 2012. Adverse birth outcomes and maternal exposure to trichloroethylene and tetrachloroethylene through soil vapor intrusion in New York State. Environ Health Perspect 120:616–621.

Formicola V, Buzhilova AP. 2004. Double child burial from Sunghir (Russia): pathology and inferences for Upper Paleolithic funerary practices. Am J Phys Anthropol 124:189–198.

Formicola V, Holt BM. 2007. Resource availability and stature decrease in Upper Palaeolithic Europe. J Anthropol Sci 85:147–155.

Fornari ED, Suszter M, Roocroft J, Bastrom T, Edmonds EW, Schlechter J. 2013. Childhood obesity as a risk factor for lateral condyle fractures over supracondylar humerus fractures. Clin Orthop Relat Res 471:1–6.

Fountain L, Krulewitch CJ. 2002. Trends in assisted reproductive technology. J Midwifery Womens Health 47:384–385.

Fox CS, Pencina MJ, Meigs JB, Vasan RS, Levitzky YS, D'Agostino RB. 2006. Trends in the incidence of type 2 diabetes mellitus from the 1970s to the 1990s: the Framingham Heart Study. Circulation 113:2914–2918.

Frayer DW, Macchiarelli R, Mussi M. 1988. A case of chondrodystrophic dwarfism in the Italian Late Upper Paleolithic. Am J Phys Anthropol 75:549–565.

Freudenheim JL, Johnson NE, Smith EL. 1986. Relationships between usual nutrient intake and bone-mineral content of women 35–65 years of age: longitudinal and cross-sectional analysis. Am J Clin Nutr 44:863–876.

Friedman JW. 2007. The prophylactic extraction of third molars: a public health hazard. Am J Public Health 97:1554–1559.

Gabay O, Hall D, Berenbaum F, Henrotin Y, Sanchez C. 2008. Osteoarthritis and obesity: experimental models. Joint Bone Spine 75:675–679.

Gadgil RM, Bhoosreddy AR, Upadhyay BR. 2012. Osteomyelitis of the mandible leading to pathological fracture in a tuberculosis patient: a case report and review of literature. Ann Trop Med Public Health 5:383–386.

Gafni RI, Baron J. 2000. Catch-up growth: possible mechanisms. Pediatr Nephrol 14:616–619.

Gaither C. 2012. Cultural conflict and the impact on non-adults at Puruchuco-Hua-querones in Peru: the case for refinement of the methods used to analyze violence against children in the archeological record. Int J Paleopath 2:69–77.

Gannagé-Yared MH, Chemali R, Yaacoub N, Halaby G. 2000. Hypovitaminosis D in a sunny country: relation to lifestyle and bone markers. J Bone Miner Res 15:1856–1862.

Garg S, Kapoor S, Malaviya AN. 2008. Diffuse idiopathic skeletal hyperostosis (DISH): an often missed diagnosis. Int J Rheum Dis 11:66–68.

Garn SM, Schwager PM. 1967. Age dynamics of persistent transverse lines in the tibia. Am J Phys Anthropol 27:375–377.

Geber J. 2012. Comparative study of perimortem weapon trauma in two early medieval skeletal populations (AD 400–1200) from Ireland. Int J Osteoarchaeol. doi:10.1002/oa.2281.

Geltman PL, Radin M, Zhang Z, Cochran J, Meyers AF. 2001. Growth status and related medical conditions among refugee children in Massachusetts, 1995–1998. Am J Public Health 91:1800–1805.

Gholipour B. 2013 Nov 20. Polio-free countries still face threat, scientists say. Live Science.

Giannecchini M, Moggi-Cecchi J. 2008. Stature in archeological samples from central Italy: methodological issues and diachronic changes. Am J Phys Anthropol 135:284–292.

Gidwani S, Fairbank A. 2004. The orthopaedic approach to managing osteoarthritis of the knee. BMJ 329:1220–1224.

Gifre L, Peris P, Monegal A, de Osaba MJM, Alvarez L, Guañabens N. 2011. Osteomalacia revisited. Clin Rheumatol 30:639–645.

Gilbert GH, Duncan RP, Crandall LA, Heft MW, Ringelberg ML. 1993. Attitudinal and behavioral characteristics of older Floridians with tooth loss. Community Dent Oral Epidemiol 21:384–389.

Gilmore CC. 2013. A comparison of antemortem tooth loss in human hunter-gatherers and non-human catarrhines: implications for the identification of behavioral evolution in the human fossil record. Am J Phys Anthropol 151:252–264.

Gładykowska-Rzeczycka JJ, Mazurek T. 2009. A rare case of forearm hypoplasia from 18th-century Gdańsk, Poland. Int J Osteoarchaeol 19:726–734.

Glencross B, Stuart-Macadam P. 2000. Childhood trauma in the archaeological record. Int J Osteoarchaeol 10:198–209.

Glencross B, Stuart-Macadam P. 2001. Radiographic clues to fractures of distal humerus in archaeological remains. Int J Osteoarchaeol 11:298–310.

Gluckman PD, Hanson MA. 2006. Evolution, development and timing of puberty. Trends Endocrinol Metab 17:7–12.

González-Reimers E, Pérez-Ramírez A, Santolaria-Fernández F, Rodríguez-Rodríguez E, Martínez-Riera A, Durán-Castellón MDC, Alemán-Valls MR, Gaspar MR. 2007. Association of Harris lines and shorter stature with ethanol consumption during growth. Alcohol 41:511–515.

Grabiner MD. 2004. Obesity and lower extremity osteoarthritis: is body mass destiny? Quest 56:41–49.

Graham GG, Adrianzen BT. 1972. Late "catch-up" growth after severe infantile malnutrition. Johns Hopkins Med J 131:204–211.

Grant P, Mata MB, Tidwell M. 2001. Femur fracture in infants: a possible accidental etiology. Pediatrics 108:1009–1011.

Green NS. 2004. Risks of birth defects and other adverse outcomes associated with assisted reproductive technology. Pediatrics 114:256–259.

Grewal J, Carmichael SL, Yang W, Shaw GM. 2012. Paternal age and congenital malformations in offspring in California, 1989–2002. Matern Child Health J 16:385–392.

Grolleau-Raoux JL, Crubezy E, Rouge D, Brugne JF, Saunders SR. 1997. Harris lines: a study of age-associated bias in counting and interpretation. Am J Phys Anthropol 103:209–217.

Guerra-Silveira F, Abad-Franch F. 2013. Sex bias in infectious disease epidemiology: patterns and processes. PLoS One 8:e62390.

Gulati D, Aggarwal AN, Kumar S, Agarwal A. 2012. Skeletal injuries following unintentional fall from height. Ulus Travma Acil Cerrahi Derg 18:141–146.

Hackshaw A, Rodeck C, Boniface S. 2011. Maternal smoking in pregnancy and birth defects: a systematic review based on 173 687 malformed cases and 11.7 million controls. Hum Reprod Update 17:589–604.

Hadjidakis DJ, Androulakis II. 2006. Bone remodeling. Ann NY Acad Sci 1092:385–396.

Haduch E, Szczepanek A, Skrzat J, Środek R, Brzegowy P. 2009. Residual rickets or osteomalacia: a case dating from the 16–18th centuries from Krosno Odrzańskie, Poland. Int J Osteoarchaeol 19:593–612.

Haidar J. 2010. Prevalence of anaemia, deficiencies of iron and folic acid and their determinants in Ethiopian women. J Health Popul Nutr 28:359–368.

Halcrow SE, Harris NJ, Tayles N, Ikehara-Quebral R, Pietrusewsky M. 2013. From the mouths of babes: dental caries in infants and children and the intensification of agriculture in mainland Southeast Asia. Am J Phys Anthropol 150:409–420.

Halpin S. 2012. Case report: the effects of massage therapy on lumbar spondylolisthesis. J Bodyw Mov Ther 16:115–123.

Hamamy H, Jamhawi L, Al-Darawsheh J, Ajlouni K. 2005. Consanguineous marriages in Jordan: why is the rate changing with time? Clin Genet 67:511–516.

Hamill J, Knutzen K. 1995. Biomechanical basis of human movement. Baltimore (MD): Williams and Wilkens.

Han SS, Baek KW, Shin MH, Kim J, Oh CS, Lee SJ, Shin DH. 2010. Dental caries prevalence of medieval Korean people. Arch Oral Biol 55:535–540.

Hanioka T, Ojima M, Tanaka K, Matsuo K, Sato F, Tanaka H. 2011. Causal assessment of smoking and tooth loss: a systematic review of observational studies. BMC Public Health 11:221.

Hannan MT, Felson DT, Pincus T. 2000. Analysis of the discordance between radiographic changes and knee pain in osteoarthritis of the knee. J Rheumatol 27:1513–1517.

Harper KN, Zuckerman MK, Harper ML, Kingston JD, Armelagos GJ. 2011. The origin and antiquity of syphilis revisited: an appraisal of Old World pre-Columbian evidence for treponemal infection. Yrbk Phys Anthropol 146:99–133.

Harris HA. 1931. Lines of arrested growth in the long bones in childhood: the correlation of histological and radiographic appearances in clinical and experimental conditions. Br J Radiol 4:534–588, 622–640.

Harris SS, Dawson-Hughes B. 1994. Caffeine and bone loss in healthy postmenopausal women. Am J Clin Nutr 60:573–578.

Harris ST, Watts NB, Genant HK, McKeever CD, Hangartner T, Keller M, Chestnut III CH, Brown J, Eriksen EF, Hoseyni MS, Axelrod DW, Miller PD. 1999. Effects of risedronate treatment on vertebral and nonvertebral fractures in women with postmenopausal osteoporosis. JAMA 282:1344–1352.

Hasegawa K, Ogose A, Morita T, Hirata Y. 2004. Painful Schmorl's node treated by lumbar interbody fusion. Spinal Cord 42:124–128.

Hayasaka K, Tomata Y, Aida J, Watanabe T, Kakizaki M, Tsuji I. 2013. Tooth loss and mortality in elderly Japanese adults: effect of oral care. J Am Geriatr Soc 61:815–820.

Hayden EC. 2014. Prenatal-screening companies expand scope of DNA tests. Nature 507:19.

Heaney RP. 2002. Effects of caffeine on bone and the calcium economy. Food Chem Toxicol 40:1263–1270.

Heaney RP. 2003. Bone mineral content, not bone mineral density, is the correct bone measure for growth studies. Am J Clin Nutr 78:350–351.

Heathcote GM, Stodder ALW, Buckley HR, Hanson DB, Douglas MT, Underwood JH, Taisipic TF, Diego VP. 1998. On treponemal disease in the Western Pacific: corrections and critique. Curr Anthropol 39:359–368.

Heino TJ, Kurata K, Higaki H, Väänänen HK. 2009. Evidence for the role of osteocytes in the initiation of targeted remodeling. Technol Health Care 17:49–56.

Hemingway CA, Parker DM, Addy M, Barbour ME. 2006. Erosion of enamel by non-carbonated soft drinks with and without toothbrushing abrasion. Br Dent J 201:447–450.

Hereen GA, Tyler J, Mandeya A. 2003. Agricultural chemical exposures and birth defects in the Eastern Cape Province, South Africa: a case-control study. Environ Health 2:11.

Hernandez M. 2013. A possible case of hypopituitarism in Neolithic China. Int J Osteoarchaeol 23:432–446.

Hernborg J, Nilsson BE. 1977. The natural course of untreated osteophytes in the knee joint, osteoarthritis, and aging. Acta Orthop Scand 44:69–74.

Hershkovitz I, Greenwald C, Rothschild BM, Latimer B, Dutour O, Jellema LM, Wish-Baratz S. 1999. Hyperostosis frontalis interna: an anthropological perspective. Am J Phys Anthropol 109:303–325.

Hewitt D, Westropp CK, Acheson RM. 1955. Oxford Child Health Survey: effect of childish ailments on skeletal development. Brit J Prev Soc Med 9:179–186.

Heymann DL, Rodier GR. 2001. Hot spots in a wired world: WHO surveillance of emerging and re-emerging infectious diseases. Lancet Infect Dis 1:345–353.

Hibbs AC, Secor WE, van Gerven D, Armelagos G. 2011. Irrigation and infection: the immunoepidemiology of schistosomiasis in ancient Nubia. Am J Phys Anthropol 145:290–298.

Hiyama Y, Yamada M, Kitagawa A, Tei N, Okada S. 2012. A four-week walking exercise programme in patients with knee osteoarthritis improves the ability of dual-task performance: a randomized controlled trial. Clin Rehabil 26:403–412.

Ho JJ, Thong MK, Nurani NK. 2006. Prenatal detection of birth defects in a Malaysian population: estimation of the influence of termination of pregnancy on birth prevalence in a developing country. Aust NZ J Obstet Gynaecol 46:55–57.

Hochberg MC, Lethbridge-Cejku M, Plato CC, Wigley FD, Tobin JD. 1991. Factors associated with OA of the hand in males, data from the Baltimore Longitudinal Study of Aging. Am J Epidemiol 134:1121–1127.

Hodges DC, Harker LA, Schermer SJ. 1990. Atresia of the external acoustic meatus in prehistoric populations. Am J Phys Anthropol 83:77–81.

Holck P. 2007. Bone mineral densities in the prehistoric, Viking-Age and medieval populations of Norway. Int J Osteoarchaeol 17:199–206.

Holland TD, O'Brien MJ. 1997. Parasites, porotic hyperostosis, and the implications of changing perspectives. Am Antiq 62:183–193.

Hollenbach KA, Barrett-Connor E, Edelstein SL, Holbrook T. 1993. Cigarette smoking and bone mineral density in older men and women. Am J Public Health 83:1265–1270.

Holmberg S, Thelin A, Thelin N. 2004. Is there an increased risk of knee osteoarthritis among farmers? A population-based case-control study. Int Arch Occup Environ Health 77:345–350.

Honein MA, Kirby RS, Meyer RE, Xing J, Skerrette NI, Yuskiv N, Marengo L, Petrini JR, Davidoff MJ, Mai CT, Druschel CM, Viner-Brown S, Sever LE. 2009. The association between major birth defects and preterm birth. Matern Child Health J 13:164–175.

Hong L, Levy SM, Warren JJ, Broffitt B. 2009. Association between enamel hypoplasia and dental caries in primary second molars: a cohort study. Caries Res 43:345–353.

Ho-Pham LT, Nguyen ND, Nguyen TV. 2009. Effect of vegetarian diets on bone mineral density: a Bayesian meta-analysis. Am J Clin Nutr 90:943–950.

Hoppe C, Mølgaard C, Michaelsen KF. 2006. Cow's milk and linear growth in industrialized and developing countries. Annu Rev Nutr 26:131–173.

Horowitz HS. 2003. The 2001 CDC recommendations for using fluoride to prevent and control dental caries in the United States. J Public Health Dent 63:3–8.

Hörup N, Melsen B, Terp S. 1987. Relationship between malocclusion and maintenance of teeth. Community Dent Oral Epidemiol 15:74–78.

Howard University, College of Sociology and Anthropology. http://www.coas.howard.edu/sociologyanthropology/anthroprogramresources.htm.

Hrdlička A. 1937. Biographical memoir of George Sumner Huntington 1861–1927. National Academy of Sciences of the United States of America Biographical Memoirs 18—eleventh memoir.

Huang JC, Peng YS, Fan JY, Jane SW, Tu LT, Chang CC, Chen MY. 2013. Factors associated with numbers of remaining teeth among type 2 diabetes: a cross-sectional study. J Clin Nurs 22:1926–1932.

Hughes C, Heylings DJA, Power C. 1996. Transverse (Harris) lines in Irish archaeological remains. Am J Phys Anthropol 101:115–131.

Hummert JR, van Gerven DP. 1985. Observations on the formation and persistence of radiopaque transverse lines. Am J Phys Anthropol 66:297–306.

Hunt IF, Murphy NJ, Henderson C, Clark VA, Jacobs RM, Johnston PK, Coulson AH. 1989. Bone mineral content in postmenopausal women: comparison of omnivores and vegetarians. Am J Clin Nutr 50:517–523.

Hunter DJ, Eckstein F. 2009. Exercise and OA. J Anat 214:197–207.

Hunter DJ, Niu J, Zhang Y, Nevitt MC, Xu L, Lui LY, Yu W, Aliabadi P, Buchanan TS, Felson DT. 2005. Knee height, knee pain, and knee osteoarthritis. Arthritis Rheum 52:1418–1423.

Hussein AS. 2011. Prevalence of intestinal parasites among school children in northern districts of West Bank-Palestine. Trop Med Int Health 16:240–244.

Hussien FH, El-Din AM, Kandeel WA, Banna RA. 2009. Spinal pathological findings in ancient Egyptians of the Greco-Roman period living in Bahriyah Oasis. Int J Osteoarchaeol 19:613–627.x

Hutchinson DL, Richman R. 2006. Regional, social, and evolutionary perspectives on treponemal infection in southeastern United States. Am J Phys Anthropol 129:544–558.

Hwang BF, Jaakkola JJ. 2003. Water chlorination and birth defects: a systematic review and meta-analysis. Arch Environ Health 58:83–91.

Hynes D, O'Brien T. 1988. Growth disturbance lines after injury of the distal tibial physis. Their significance in prognosis. Bone Joint J Br 70:231–233.

Ibáñez L, Jiménez R, de Zegher F. 2006. Early puberty-menarche after precocious pubarche: relation to prenatal growth. Pediatrics 117:117–121.

Indahl A. 2004. Low back pain: diagnosis, treatment, and prognosis. Scand J Rheumatol 33:199–209.

Inoue K, Hukuda S, Nakai M, Katayama K, Huang J. 1999. Erosive peripheral polyarthritis in ancient Japanese skeletons: a possible case of rheumatoid arthritis. Int J Osteoarchaeol 9:1–7.

Irving RJ, Belton NR, Elton RA, Walker BR. 2000. Adult cardiovascular risk factors in premature babies. Lancet 355:2135–2136.

Iwamoto J, Abe H, Tsukimura Y, Wakano K. 2005. Relationship between radiographic abnormalities of lumbar spine and incidence of low back pain in high school rugby players: a prospective study. Scand J Med Sci Sports 15:163–168.

Jabbar Z, Aggarwal PK, Chandel N, Kohli HS, Gupta KL, Sakhuja V, Jha V. 2009. High prevalence of vitamin D deficiency in north Indian adults is exacerbated in those with chronic kidney disease. Nephrology 14:345–349.

Jackes MK. 1983. Osteological evidence for smallpox: a possible case from seventeenth century Ontario. Am J Phys Anthropol 60:75–81.

Jaleel R, Nasrullah FD, Khan A. 2010. Osteopenia in younger females. J Surg Pakistan 15:29–33.

Jankauskas R, Urbanavičius A. 2008. Possible indications of metabolic syndrome in Lithuanian paleoosteological materials. Papers on Anthropology 17:103–112.

Jansson L, Lavstedt S, Zimmerman M. 2002. Prediction of marginal bone loss and tooth loss—a prospective study over 20 years. J Clin Periodontol 29:672–678.

Jauniaux E, Ben-Ami I, Maymon R. 2013. Do assisted-reproduction twin pregnancies require additional antenatal care? Reprod Biomed Online 26:107–119.

Jensen MC, Brant-Zawadzki MN, Obuchowski N, Modic MT, Malkasian D, Ross JS. 1994. Magnetic resonance imaging of the lumbar spine in people without back pain. N Engl J Med 331:69–73.

Jette AM, Feldman HA, Tennstedt SL. 1993. Tobacco use: a modifiable risk factor for dental disease among the elderly. Am J Public Health 83:1271–1276.

Jiang LS, Zhang ZM, Jiang SD, Chen WH, Dai LY. 2008. Differential bone metabolism between postmenopausal women with OA and osteoporosis. J Bone Miner Res 23:475–483.

Jimenez M, Dietrich T, Shih MC, Li Y, Joshipura KJ. 2009. Racial/ethnic variations in associations between socioeconomic factors and tooth loss. Community Dent Oral Epidemiol 37:267–275.

Jimenez M, Hu FB, Marino M, Li Y, Joshipura KJ. 2012. Type 2 diabetes mellitus and 20 year incidence of periodontitis and tooth loss. Diabetes Res Clin Pract 98:494–500.

Jiménez-Brobeil SA, Al Oumaoui I, Du Souich PH. 2007. Childhood trauma in several populations from the Iberian Peninsula. Int J Osteoarchaeol 17:189–198.

Jiménez-Brobeil SA, Al Oumaoui I, Du Souich P. 2010. Some types of vertebral pathologies in the Argar culture (Bronze Age, SE Spain). Int J Osteoarchaeol 20:36–46.

Johansson I, Kressin NR, Nunn ME, Tanner AC. 2010. Snacking habits and caries in young children. Caries Res 44:421–430.

Jokar M, Hatef Fard M, Mirfeizi Z. 2008. Rickets and osteomalacia in northeast Iran: report of 797 cases. Int J Rheum Dis 11:170–174.

Jones CM, Worthington H. 1999. Fluoridation: the relationship between water fluoridation and socioeconomic deprivation on tooth decay in 5-year-old children. Br Dent J 186:397–400.

Jones K, Smith D. 1973. Recognition of the fetal alcohol syndrome in early infancy. Lancet 302:999–1001.

Jonsson H, Manolescu I, Stefansson SE, Ingvarsson T, Jonsson HH, Manolescu A, Gulcher J, Stefansson K. 2003. The inheritance of hand osteoarthritis in Iceland. Arthritis Rheum 48:391–395.

Jorde LB, Fineman RM, Martin RA. 1983. Epidemiology and genetics of neural tube defects: an application of the Utah Genealogical Data Base. Am J Phys Anthropol 62:23–31.

Jurmain RD. 1977. Stress and the etiology of osteoarthritis. Am J Phys Anthropol 46:353–365.

Jurmain RD. 1991. Degenerative changes in peripheral joints as indicators of mechanical stress: opportunities and limitations. Int J Osteoarchaeol 1:247–252.

Jurmain RD. 1999. Stories from the skeleton: behavioral reconstruction in human osteology. London: Taylor and Francis.

Kacki S, Duneufjardin P, Blanchard P, Castex D. 2013. Humerus varus in a subadult skeleton from the medieval graveyard of La Madeleine (Orléans, France). Int J Osteoarchaeol 23:119–126.

Kahl KE, Smith MO. 2000. The pattern of spondylolysis deformans in prehistoric samples from west-central New Mexico. Int J Osteoarchaeol 10:432–446.

Kalichman L, Kim DH, Li L, Guermazi A, Berkin V, Hunter DJ. 2009. Spondylolysis and spondylolisthesis: prevalence and association with low back pain in the adult community-based population. Spine 34:199–205.

Kalichman L, Korosteshevsky M, Batsevich V, Kobyliansky E. 2011. Climate is associated with prevalence and severity of radiographic hand osteoarthritis. HOMO 62:280–287.

Kalpakcioglu B, Altınbilek T, Senel K. 2009. Determination of spondylolisthesis in low back pain by clinical evaluation. J Back Musculoskelet Rehabil 22:27–32.

Kanis JA. 1981. Osteomalacia and chronic renal failure. J Clin Pathol 34:1295–1307.

Kanis JA, Johansson H, Johnell O, Oden A, De Laet C, Eisman JA, Pols H, Tenenhouse A. 2005. Alcohol intake as a risk factor for fracture. Osteoporosis Int 16:737–742.

Kaplowitz PB, Oberfield SE. 1999. Reexamination of the age limit for defining when puberty is precocious in girls in the United States: implications for evaluation and treatment. Pediatrics 104:936–941.

Katz JN, Brophy RH, Chaisson CE, de Chaves L, Cole BJ, Dahm DL, Donnell-Fink LA, Guermazi A, Haas AK, Jones MH, et al. 2013. Surgery versus physical therapy for a meniscal tear and osteoarthritis. N Engl J Med 368:1675–1684.

Kaye EK, Valencia A, Baba N, Spiro A, Dietrich T, Garcia RI. 2010. Tooth loss and periodontal disease predict poor cognitive function in older men. J Am Geriatr Soc 58:713–718.

Kazaura M, Lie RT, Skjærven R. 2004. Paternal age and the risk of birth defects in Norway. Ann Epidemiol 14:566–570.

Kazaura MR, Lie RT. 2002. Down's syndrome and paternal age in Norway. Paediatr Perinat Epidemiol 16:314–319.

Keaveny TM, Morgan EF, Niebur GL, Yeh OC. 2001. Biomechanics of trabecular bone. Annu Rev Biomed Eng 3:307–333.

Keays G, Swaine B, Ehrmann-Feldman D. 2006. Association between severity of musculoskeletal injury and risk of subsequent injury in children and adolescents on the basis of parental recall. Arch Pediatr Adolesc Med 160:812–816.

Keenleyside A. 2008. Dental pathology and diet at Apollonia, a Greek colony on the Black Sea. Int J Osteoarchaeol 18:262–279.

Keenleyside A. 2012. Sagittal clefting of the fifth lumbar vertebra of a young adult female from Apollonia Pontica, Bulgaria. Int J Osteoarchaeol. doi:10.1002/oa.2268.

Keizer-Schrama SMPF, Mul D. 2001. Trends in pubertal development in Europe. Apmis 109:S164–S170.

Kelley MA. 1982. Intervertebral osteochondrosis in ancient and modern populations. Am J Phys Anthropol 59:271–279.

Kelley MA, El-Najjar MY. 1980. Natural variation and differential diagnosis of skeletal changes in tuberculosis. Am J Phys Anthropol 52:153–167.

Kelley-Quon LI, Tseng CH, Janzen C, Shew SB. 2013. Congenital malformations associated with assisted reproductive technology: a California statewide analysis. J Pediatr Surg 48:1218–1224.

Kemkes-Grottenthaler A. 2005. The short die young: the interrelationship between stature and longevity—evidence from skeletal remains. Am J Phys Anthropol 128:340–347.

Kemp AM, Dunstan F, Harrison S, Morris S, Mann M, Rolfe K, Datta S, Thomas DP, Sibert JR, Maguire S. 2008. Patterns of skeletal fractures in child abuse: systematic review. BMJ 337.

Keyes KM, Li G, Hasin DS. 2011. Birth cohort effects and gender differences in alcohol epidemiology: a review and synthesis. Alcohol Clin Exp Res 35:2101–2112.

Khlat M. 1989. Inbreeding effects on fetal growth in Beirut, Lebanon. Am J Phys Anthropol 80:481–484.

Khoshnood B, Pryde P, Wall S, Singh J, Mittendorf R, Lee KS. 2000. Ethnic differences in the impact of advanced maternal age on birth prevalence of Down syndrome. Am J Public Health 90:1778–1781.

Kilgore L. 1989. Possible case of rheumatoid arthritis from Sudanese Nubia. Am J Phys Anthropol 79:177–183.

Kilgore L. 1990. Biomechanical relationship in the development of degenerative joint disease of the spine. Paper presentation at: Eighth European Meeting of the Paleopathology Association; Cambridge, England.

Kim JH, Scialli AR. 2011. Thalidomide: the tragedy of birth defects and the effective treatment of disease. Toxicol Sci 122:1–6.

Kim MJ, Lee IS, Kim YS, Oh CS, Park JB, Shin MH, Shin DH. 2012. Diffuse idiopathic skeletal hyperostosis cases found in Joseon Dynasty human sample collection of Korea. Int J Osteoarchaeol 22:235–244.

Kimura K. 1984. Studies on growth and development in Japan. Yrbk Phys Anthropol 27:179–213.

King J, Diefendorf D, Apthorp J, Negrete VF, Carlson M. 1988. Analysis of 429 fractures in 189 battered children. J Pediatr Orthop 8:585–589.

Kiss C, Szilagyi M, Paksy A, Poor G. 2002. Risk factors for diffuse idiopathic skeletal hyperostosis: a case-control study. Rheumatology 41:27–30.

Kjellström A. 2012. Possible cases of leprosy and tuberculosis in medieval Sigtuna, Sweden. Int J Osteoarchaeol 22:261–283.

Kleinman PK, Marks Jr SC, Richmond JM, Blackbourne BD. 1995. Inflicted skeletal injury: a postmortem radiologic-histopathologic study in 31 infants. AJR Am J Roentgenol 165:647–650.

Kleinman PK, Nimkin K, Spevak MR, Rayder SM, Madansky DL, Shelton YA, Patterson MM. 1996. Follow-up skeletal surveys in suspected child abuse. AJR Am J Roentgenol 167:893–896.

Klingele KE, Kocher MS. 2002. Little league elbow. Sports Med 32:1005–1015.

Knops NB, Sneeuw KC, Brand R, Hille ET, den Ouden AL, Wit JM, Verloove-Vanhorick SP. 2005. Catch-up growth up to ten years of age in children born very preterm or with very low birth weight. BMC Pediatrics 5:26.

Knutsson S, Engberg IB. 1999. An evaluation of patients' quality of life before, 6 weeks and 6 months after total hip replacement surgery. J Adv Nurs 30:1349–1359.

Kocher MS, Waters PM, Micheli LJ. 2000. Upper extremity injuries in the paediatric athlete. Sports Med 30:117–135.

Kojima N, Douchi T, Kosha S, Nagata Y. 2002. Cross-sectional study of the effects of parturition and lactation on bone mineral density later in life. Maturitas 41:203–209.

Kolar JC, Munro IR, Farkas LG. 1987. Anthropometric evaluation of dysmorphology in craniofacial anomalies: Treacher Collins syndrome. Am J Phys Anthropol 74:441–451.

Kolker S, Itsekzon T, Yinnon AM, Lachish T. 2012. Osteomyelitis due to *Salmonella enterica subsp. arizonae*: the price of exotic pets. Clin Microbiol Infect 18:167–170.

Komlos J, Breitfelder A. 2007. Are Americans shorter (partly) because they are fatter? A comparison of US and Dutch children's height and BMI values. Ann Hum Biol 34:593–606.

Komlos J, Breitfelder A. 2008. Differences in the physical growth of US-born black and white children and adolescents ages 2–19, born 1942–2002. Ann Hum Biol 35:11–21.

Korpelainen R, Keinanen-Kiukaanniemi S, Nieminen P, Heikkinen J, Vaananen K, Korpelainen J. 2010. Long-term outcomes of exercise: follow-up of a randomized trial in older women with osteopenia. Arch Intern Med 170:1548–1556.

Kozenko MM, Chudley AE. 2010. Genetic implications and health consequences following the Chernobyl nuclear accident. Clin Genet 77:221–226.

Kozieradzka-Ogunmakin I. 2011. Multiple epiphyseal dysplasia in an Old Kingdom Egyptian skeleton: a case report. Int J Paleopath 1:200–206.

Krishan G. 1986. Effect of parental consanguinity on anthropometric measurements among the Sheikh Sunni Muslim boys of Delhi. Am J Phys Anthropol 70:69–73.

Kristjansdottir G, Rhee H. 2002. Risk factors of back pain frequency in schoolchildren: a search for explanations to a public health problem. Acta Paediatr 91:849–854.

Kritz-Silverstein D, Barrett-Connor E. 1993. Bone mineral density in postmenopausal women as determined by prior oral contraceptive use. Am J Public Health 83:100–102.

Krølner B, Nielsen SP. 1982. Bone mineral content of the lumbar spine in normal and osteoporotic women: cross-sectional and longitudinal studies. Clin Sci 62:329–336.

Kutterer A, Alt KW. 2008. Cranial deformations in an Iron Age population from Münsingen-Rain, Switzerland. Int J Osteoarchaeol 18:392–406.

Kuzma C. 2012. Overdoing it: more young athletes are getting major-league injuries. Current Health Teens 38:10–12.

Kyere KA, Than KD, Wang AC, Rahman SU, Valdivia-Valdivia JM, La Marca F, Park P. 2012. Schmorl's nodes. Eur Spine J 21:2115–2121.

Lalumandier JA, Ayers LW. 2000. Fluoride and bacterial content of bottled water vs tap water. Arch Fam Med 9:246–250.

Lambert PM. 2002. Rib lesions in a prehistoric Puebloan sample from southwestern Colorado. Am J Phys Anthropol 117:281–292.

Lanfranco LP, Eggers S. 2010. The usefulness of caries frequency, depth, and location in determining cariogenicity and past subsistence: a test on early and later agriculturalists from the Peruvian coast. Am J Phys Anthropol 143:75–91.

Lanyon P, Muir K, Doherty S, Doherty M. 2000. Assessment of a genetic contribution to osteoarthritis of the hip: sibling study. BMJ 321:1179–1183.

Laor T, Jaramillo D. 2009. MR imaging insights into skeletal maturation: what is normal? Radiology 250:28–38.

Larsen CS. 1987. Bioarchaeological interpretations of subsistence economy and behavior from human skeletal remains. Advances in Archaeological Method and Theory 10:339–445.

Larsen CS. 1995. Biological changes in human populations with agriculture. Annu Rev Anthropol 24:185–213.

Larsen CS. 1997. Bioarchaeology: Interpreting behavior from the human skeleton. London: Cambridge.

Lavelle CLB. 1976. A study of multiracial malocclusions. Community Dent Oral Epidemiol 4:38–41.

Law MR, Hackshaw AK. 1997. A meta-analysis of cigarette smoking, bone mineral density and risk of hip fracture: recognition of a major effect. BMJ 315:841–846.

Lechner M, Steirer I, Brinkhaus B, Chen Y, Krist-Dungl C, Koschier A, Gantschacher M, Neumann K, Zauner-Dungl A. 2011. Efficacy of individualized Chinese herbal medication in osteoarthrosis of hip and knee: a double-blind, randomized-controlled clinical study. J Altern Complement Med 17:539–547.

Lee J, Song J, Hootman JM, Semanik PA, Chang RW, Sharma L, van Horn L, Bathon JM, Eaton CM, Hochberg MC, et al. 2013. Obesity and other modifiable factors for physical inactivity measured by accelerometer in adults with knee osteoarthritis. Arthritis Care Res 65:53–61.

Lee MJ, Riew KD. 2009. The prevalence cervical facet arthrosis: an osseous study in a cadveric population. Spine J 9:711–714.

Lees B, Stevenson JC, Molleson T, Arnett TR. 1993. Differences in proximal femur bone density over two centuries. Lancet 341:673–676.

Leitch A. 2012 Jan 24. E-mail message to author.

Leonard MB, Shults J, Wilson BA, Tershakovec AM, Zemel BS. 2004. Obesity during childhood and adolescence augments bone mass and bone dimensions. Am J Clin Nutr 80:514–523.

Leonard WR, Spencer GJ, Galloway VA, Osipova L. 2002. Declining growth status of indigenous Siberian children in post-Soviet Russia. Hum Biol 74:197–209.

Lethbridge-Çejku M, Scott WW, Reichle R, Ettinger WH, Zonderman A, Costa P, Plato CC, Tobin JC, Hochberg MC. 1995. Association of radiographic features of osteoarthritis of the knee with knee pain: data from the Baltimore Longitudinal Study of Aging. Arthritis Rheum 8:182–188.

Lewiecki EM, Urig EJ, Williams RC. 2008. Tumor-induced osteomalacia: lessons learned. Arthritis Rheum 58:773–777.

Lewis ME. 2010. Life and death in a civitas capital: metabolic disease and trauma in the children from late Roman Dorchester, Dorset. Am J Phys Anthropol 142:405–416.

Lewis ME. 2011. Tuberculosis in the non-adults from Romano-British Poundbury Camp, Dorset, England. Int J Paleopath 1:12–23.

Liebschner MA. 2004. Biomechanical considerations of animal models used in tissue engineering of bone. Biomaterials 25:1697–1714.

Lieverse AR, Weber AW, Bazaliiskiy VI, Goriunova OI, Savel'ev NA. 2007. Osteoarthritis in Siberia's Cis-Baikal: skeletal indicators of hunter-gatherer adaptation and cultural change. Am J Phys Anthropol 132:1–16.

Lim CED, Cheng NCL. 2011. Obesity and reproduction. Journal of the Australian Traditional-Medicine Society 17:143–145.

Liu M, Zhu L, Zhang B, Petersen PE. 2007. Changing use and knowledge of fluoride toothpaste by schoolchildren, parents and schoolteachers in Beijing, China. Int Dent J 57:187–194.

Lloyd T, Rollings N, Eggli DF, Kieselhorst K, Chinchilli VM. 1997. Dietary caffeine intake and bone status of postmenopausal women. Am J Clin Nutr 65:1826–1830.

Lomenick JP, Calafat AM, Melguizo Castro MS, Mier R, Stenger P, Foster MB, Wintergerst KA. 2010. Phthalate exposure and precocious puberty in females. J Pediatr 156:221–225.

Losina E, Walensky RP, Reichmann WM, Holt HL, Gerlovin H, Solomon DH, Jordan JM, Hunter DJ, Suter LG, Weinstein AM, et al. 2011. Impact of obesity and knee osteoarthritis on morbidity and mortality in older Americans. Ann Intern Med 154:217–226.

Lovejoy CO, Heiple KG. 1981. The analysis of fractures in skeletal populations with an example from the Libben site, Ottowa County, Ohio. Am J Phys Anthropol 55:529–541.

Lowe G, Woodward M, Rumley A, Morrison C, Tunstall-Pedoe H, Stephen K. 2003. Total tooth loss and prevalent cardiovascular disease in men and women: possible roles of citrus fruit consumption, vitamin C, and inflammatory and thrombotic variables. J Clin Epidemiol 56:694–700.

Lupo PJ, Canfield MA, Chapa C, Lu W, Agopian AJ, Mitchell LE, Shaw GM, Waller DK, Olshan AF, Finnell RH, Zhu H. 2012. Diabetes and obesity-related genes and the risk of neural tube defects in the National Birth Defects Prevention Study. Am J Epidemiol 176:1101–1109.

Macallan DC, Maxwell JD, Eastwood JB. 1992. Osteomalacia should be sought and treated before withdrawal of anticonvulsant therapy in UK Asians. Postgrad Med J 68:134–136.

MacFarlane G, de Silva V, Jones G. 2011. The relationship between body mass index across the life course and knee pain in adulthood: results from the 1958 birth cohort study. Rheumatology 50:2251–2256.

Mackowiak PA, Blos VT, Aguilar M, Buikstra JE. 2005. On the origin of American tuberculosis. Clin Infect Dis 41:515–518.

Madimenos FC. 2011. Reproductive trade-offs in skeletal health and physical activity among the indigenous Shuar of Ecuadorian Amazonia: a life history approach [PhD dissertation]. University of Oregon.

Magalhães RJS, Clements AC. 2011. Mapping the risk of anaemia in preschool-age children: the contribution of malnutrition, malaria, and helminth infections in West Africa. PLoS Med 8:e1000438.

Mahoney P. 2006. Dental microwear from Natufian hunter-gatherers and early Neolithic farmers: comparisons within and between samples. Am J Phys Anthropol 130:308–319.

Malleson P, Clinch J. 2003. Pain syndromes in children. Curr Opin Rheumatol 15:572–580.

Mamiro PS, Kolsteren P, Roberfroid D, Tatala S, Opsomer AS, van Camp JH. 2005. Feeding practices and factors contributing to wasting, stunting, and iron-deficiency anaemia among 3–23-month-old children in Kilosa district, rural Tanzania. J Health Popul Nutr 23:222–230.

Mandelstam SA, Cook D, Fitzgerald M, Ditchfield MR. 2003. Complementary use of radiological skeletal survey and bone scintigraphy in detection of bony injuries in suspected child abuse. Arch Dis Child 88:387–390.

Manek NJ, Hart D, Spector TD, MacGregor AJ. 2003. The association of body mass index and osteoarthritis of the knee joint. Arthritis Rheum 48:1024–1020.

Mangano JJ, Gould JM, Sternglass EJ, Sherman JD, Brown J, McDonnell W. 2002. Infant death and childhood cancer reductions after nuclear plant closings in the United States. Arch Environ Health 57:23–31.

Mann RW, Thomas MD, Adams BJ. 1998. Congenital absence of the ulna with humeroradial synostosis in a prehistoric skeletal from Moundville, Alabama. Int J Osteoarchaeol 8:295–299.

Marcelli C, Favier F, Kotzki PO, Ferrazzi V, Picot M-C, Simon L. 1995. The relationship between OA of the hands, bone mineral density, and osteoporotic fractures in elderly women. Osteoporos Int 5:382–388.

March of Dimes. http://www.marchofdimes.org.

Marini JC. 2003. Do bisphosphonates make children's bones better or brittle? N Engl J Med 349:423–426.

Márquez-Grant N. 2009. Caries correction factors applied to a Punic (6th–2nd BC) population from Ibiza (Spain). Bull Int Assoc Paleodont 3:20–29.

Marsh AG, Sanchez TV, Michelsen O, Chaffee FL, Fagal SM. 1988. Vegetarian lifestyle and bone mineral density. Am J Clin Nutr 48:837–841.

Marshall WA, Tanner JM. 1969. Variations in pattern of pubertal changes in girls. Arch Dis Child 44:291–303.

Martin DC, Danforth ME. 2009. An analysis of secular change in the human mandible over the last century. Am J Human Biol 21:704–706.

Martin MV, Kanatas AN, Hardy P. 2005. Antibiotic prophylaxis and third molar surgery. Br Dent J 198:327–330.

Martinez H, Brunson EK. 2012. I don't need it, you can have it: motivations for whole body donation to FACTS. American Anthropological Association Annual Meeting.

Martínez-Lavin M, Mansilla J, Pineda C, Pijoan C. 1995. Ankylosing spondylitis is indigenous to Mesoamerica. J Rheumatol 22:2327–2330.

Masharawi Y, Dar G, Peleg S, Steinberg N, Alperovitch-Najenson D, Salame K, Hershkovitz I. 2007. Lumbar facet anatomy changes in spondylolysis: a comparative skeletal study. Eur Spine J 16:993–999.

Masharawi Y, Salame K, Mirovsky Y, Peleg S, Dar G, Steinberg N, Hershkovitz I. 2008. Vertebral body shape variation in the thoracic and lumbar spine: characterization of its asymmetry and wedging. Clin Anat 21:46–54.

Masnicová S, Beňus R. 2001. Atresia of an external acoustic meatus in an individual from historical Bratislava (Slovakia). Anthropol Sci 109:315–324.

Masnicová S, Beňus R. 2003. Developmental anomalies in skeletal remains from the Great Morovia and Middle Ages cemeteries at Devín (Slovakia). Int J Osteoarchaeol 13:266–274.

Materna-Kiryluk A, Wiśniewska K, Badura-Stronka M, Mejnartowicz J, Więckowska B, Balcar-Boroń A, Czerwionka-Szaflarska M, Gajewska E, Godula-Stuglik U, Krawczyński M, et al. 2009. Parental age as a risk factor for isolated congenital malformations in a Polish population. Paediatr Perinat Epidemiol 23:29–40.

Mathew AJ, Nair JB, Pillai SS. 2011. Rheumatic-musculoskeletal manifestations in type 2 diabetes mellitus patients in south India. Int J Rheum Dis 14:55–60.

Mathuram A, Rijn RV, Varghese GM. 2013. Salmonella typhi rib osteomyelitis with abscess mimicking a "cold abscess." J Glob Infect Dis 5:80.

Matkovic V, Jelic T, Wardlaw GM, Ilich JZ, Goel PK, Wright JK, Andon MB, Smith KT, Heaney RP. 1994. Timing of peak bone mass in Caucasian females and its implication for the prevention of osteoporosis. Inference from a cross-sectional model. J Clin Invest 93:799–808.

Matos V, Santos AL. 2006. On the trail of pulmonary tuberculosis based on rib lesions: from the human identified skeletal collection from Museu Bocage (Lisbon, Portugal). Am J Phys Anthropol 130:190–200.

Mavroudas SR. 2012 Nov 29. E-mail message to author.

May H, Peled N, Dar G, Abbas J, Hershkovitz I. 2011. Hyperostosis frontalis interna: what does it tell us about our health? Am J Hum Biol 23:392–397.

Mayer JD. 2000. Geography, ecology and emerging infectious diseases. Soc Sci Med 50:937–952.

Mays SA. 1995. The relationship between Harris lines and other aspects of skeletal development in adults and juveniles. J Archaeol Sci 22:511–520.

Mays SA. 2002. The relationship between molar wear and age in an early 19th century AD archaeological human skeletal series of documented age at death. J Archaeol Sci 29:861–871.

Mays SA. 2006a. Age-related cortical bone loss in women from a 3rd–4th century AD population from England. Am J Phys Anthropol 129:518–528.

Mays SA. 2006b. Spondylolysis, spondylolisthesis, and lumbo-sacral morphology in a medieval English skeletal population. Am J Phys Anthropol 131:352–362.

Mays SA. 2007. Lysis at the anterior vertebral body margin: evidence for brucellar spondylitis? Int J Osteoarchaeol 17:107–118.

Mays SA, Brickley M, Ives R. 2009. Growth and vitamin D deficiency in a population from 19th century Birmingham, England. Int J Osteoarchaeol 19:406–415.

Mays SA, Lees B, Stevenson JC. 1998. Age-dependent bone loss in the femur in a medieval population. Int J Osteoarchaeol 8:97–106.

Mays SA, Taylor GM. 2003. A first prehistoric case of tuberculosis from Britain. Int J Osteoarchaeol 13:189–196.

Mays SA, Turner-Walker G, Syversen U. 2006. Osteoporosis in a population from medieval Norway. Am J Phys Anthropol 131:343–351.

Mays SA, Vincent S, Meadows J. 2012. A possible case of treponemal disease from England dating to the 11th–12th century AD. Int J Osteoarchaeol 22:366–372.

Mazess RB, Barden HS. 1991. Bone density in premenopausal women: effects of age, dietary intake, physical activity, smoking, and birth-control pills. Am J Clin Nutr 53:132–142.

Mazess RB, Mather W. 1974. Bone mineral content of north Alaskan Eskimos. Am J Clin Nutr 27:916–925.

McIntyre MH. 2011. Adult stature, body proportions and age at menarche in the United States National Health and Nutrition Examination Survey (NHANES) III. Ann Hum Biol 38:716–720.

McKeever A. 2012. Genetics versus environment in the aetiology of malocclusion. Br Dent J 212:527–528.

McMorland G, Suter E. 2000. Chiropractic management of mechanical neck and low-back pain: a retrospective, outcome-based analysis. J Manipulative Physiol Ther 23:307–311.

Meadows Jantz L, Jantz RL. 1999. Secular change in long bone length and proportion in the United States, 1800–1970. Am J Phys Anthropol 110:57–67.

Megson E, Kapellas K, Bartold PM. 2010. Evidence synthesis: Relationship between periodontal disease and osteoporosis. Int J Evid Based Healthc 8:129–139.

Meling T, Harboe K, Enoksen CH, Aarflot M, Arthursson AJ, Søreide K. 2013. Reliable classification of children's fractures according to the comprehensive classification of long bone fractures by Müller. Acta Orthop 84:207–212.

Meller C, Urzua I, Moncada G, von Ohle C. 2009. Prevalence of oral pathologic findings in an ancient pre-Columbian archeologic site in the Atacama Desert. Oral Dis 15:287–294.

Mendes E. 2011. In U.S., self-reported weight up nearly 20 pounds since 1990. [accessed February 24, 2014]. http://www.gallup.com/poll/150947/self-reported-weight-nearly-pounds-1990.aspx.

Mensforth RP, Latimer BM. 1989. Hamann-Todd collection aging studies: osteoporosis fracture syndrome. Am J Phys Anthropol 80:461–479.

Merbs CF. 1983. Patterns of activity-induced pathology in a Canadian Inuit population. Archaeological Survey of Canada, Mercury Series Paper, 119.

Merbs CF. 1992. A new world of infectious disease. Yrbk Phys Anthropol 35:3–42.

Merbs CF. 1995. Incomplete spondylolysis and healing: a study of ancient Canadian Eskimo skeletons. Spine 20:2328–2334.

Merbs CF. 1996. Spondylolysis and spondylolisthesis: a cost of being an erect biped or a clever adaptation. Yrbk Phys Anthropol 39:201–228.

Merbs CF. 2001. Degenerative spondylolisthesis in ancient and historic skeletons from New Mexico Pueblo sites. Am J Phys Anthropol 116:285–295.

Merbs CF. 2002. Spondylolysis in Inuit skeletons from Arctic Canada. Int J Osteoarchaeol 12:279–290.

Merbs CF, Euler RC. 1985. Atlanto-occipital fusion and spondylolisthesis in an Anasazi skeleton from Bright Angel Ruin, Grand Canyon National Park, Arizona. Am J Phys Anthropol 67:381–391.

Meredith HV. 1941. Stature and weight of private school children in two successive decades. Am J Phys Anthropol 28:1–40.

Merkle D, McDonald DD. 2009. Use of recommended osteoarthritis pain treatment by older adults. J Adv Nurs 65:828–835.

Meyer C, Jung C, Kohl T, Poenicke A, Poppe A, Alt KW. 2002. Syphilis 2001—a palaeopathological reappraisal. Homo 53:39–58.

Michaëlsson K, Baron JA, Farahmand BY, Ljunghall S. 2001. Influence of parity and lactation on hip fracture risk. Am J Epidemiol 153:1166–1172.

Miller E, Ragsdale BD, Ortner DJ. 1996. Accuracy in dry bone diagnosis: a comment on palaeopathological methods. Int J Osteoarchaeol 6:221–229.

Miller SM, Kukuljan S, Turner AI, van der Pligt P, Ducher G. 2012. Energy deficiency, menstrual disturbances, and low bone mass: what do exercising Australian women know about the female athlete triad? Int J Sport Nutr Exerc Metab 22:131–138.

Minozzi S, Manzi G, Ricci F, di Lernia S, Borgognin Tarli SM. 2003. Nonalimentary tooth use in prehistory: an example from early Holocene in central Sahara (Uan Muhuggiag, Tadrart Acacus, Libya). Am J Phys Anthropol 120:225–232.

Mitchell LE, Adzick NS, Melchionne J, Pasquariello PS, Sutton LN, Whitehead AS. 2004. Spina bifida. Lancet 364:1885–1895.

Mitchell P. 2003. The archaeological study of epidemic and infectious disease. World Archaeol 35:171–179.

Mølgaard C, Thomsen BL, Prentice A, Cole TJ, Michaelsen KF. 1997. Whole body bone mineral content in healthy children and adolescents. Arch Dis Child 76:9–15.

Molnar P. 2008. Dental wear and oral pathology: possible evidence and consequences of habitual use of teeth in a Swedish Neolithic sample. Am J Phys Anthropol 136:423–431.

Morris FL, Naughton GA, Gibbs JL, Carlson JS, Wark JD. 1997. Prospective ten-month exercise intervention in premenarcheal girls: positive effects on bone and lean mass. J Bone Miner Res 12:1453–1462.

Moscowitz RW. 1993. Clinical and laboratory findings in osteoarthritis. In: McCarty DJ, Coopman WJ, editors. Arthritis and Allied Conditions. Philadelphia (PA): Lea and Febinger. p. 1735–1760.

Mosley JR. 2000. Osteoporosis and bone functional adaptation: mechanobiological regulation of bone architecture in growing and adult bone, a review. J Rehabil Res Dev 37:189–199.

Moss ME, Lanphear BP, Auinger P. 1999. Association of dental caries and blood lead levels. JAMA 281:2294–2298.

Mulhern DM, Wilczak CA. 2012. Frequency of complete cleft sacra in a Native American sample. Int J Osteoarchaeol. doi:10.1002/oa.2280.

Müller F, Naharro M, Carlsson GE. 2007. What are the prevalence and incidence of tooth loss in the adult and elderly population in Europe? Clin Oral Implants Res 18:2–14.

Mummert A, Esche E, Robinson J, Armelagos GJ. 2011. Stature and robusticity during the agricultural transition: evidence from the bioarchaeological record. Econ Hum Biol 9:284–301.

Munns CF, Simm PJ, Rodda CP, Garnett SP, Zacharin MR, Ward LM, Geddes J, Cherian S, Zurynski Y, Cowell C. 2012. Incidence of vitamin D deficiency rickets among Australian children: an Australian Paediatric Surveillance Unit study. Med J Aust 196:466–468.

Murphy EM. 1996. A possible case of hydrocephalus in a medieval child from Doonbought Fort, Co. Antrim, Northern Ireland. Int J Osteoarchaeol 6:435–442.

Murphy EM, Chistov YK, Hopkins R, Rutland P, Taylor GM. 2009. Tuberculosis among Iron Age individuals from Tyva, South Siberia: palaeopathological and biomolecular findings. J Archaeol Sci 36:2029–2038.

Must A, Strauss RS. 1999. Risks and consequences of childhood and adolescent obesity. Int J Obes Relat Metab Disord 23:2–11.

Mutolo MJ, Jenny LL, Buszek AR, Fenton TW, Foran DR. 2012. Osteological and molecular identification of brucellosis in ancient Butrint, Albania. Am J Phys Anthropol 147:254–263.

National Institutes of Health. http://www.nlm.nih.gov.

Ndreu D. 2006. Keeping bad science out of the courtroom: why post-Daubert courts are correct in excluding opinions based on animal studies from birth-defects cases. Gold Gate Univ Law Rev 36:459–488.

Neal B, Gray H, MacMahon S, Dunn L. 2002. Incidence of heterotopic bone formation after major hip surgery. ANZ J Surg 72:808–821.

Neidell M, Herzog K, Glied S. 2010. The association between community water fluoridation and adult tooth loss. Am J Public Health 100:1980–1985.

Nelson DA, Barondess DA, Hendrix SL, Beck TJ. 2000. Cross-sectional geometry, bone strength, and bone mass in the proximal femur in black and white postmenopausal women. J Bone Miner Res 15:1992–1997.

Nelson GC, Lukacs JB, Yule P. 1999. Dates, caries, and early tooth loss during the Iron Age of Oman. Am J Phys Anthropol 108:333–343.

Neve A, Corrado A, Cantatore FP. 2012. Osteocytes: central conductors of bone biology in normal and pathological conditions. Acta Physiol (Oxf) 204:317–330.

Nevitt MC, Xu L, Zhang Y, Lui LY, Yu W, Lane NE, Qin M, Hocberg MC, Cummings SR, Felson DT. 2002. Very low prevalence of hip osteoarthritis among Chinese elderly in Beijing, China, compared with whites in the United States: the Beijing osteoarthritis study. Arthritis Rheum 46:1773–1779.

New SA, Robins SP, Campbell MK, Martin JC, Garton MJ, Bolton-Smith C, Grubb DA, Lee SJ, Reid DM. 2000. Dietary influences on bone mass and bone metabolism: further evidence of a positive link between fruit and vegetable consumption and bone health? Am J Clin Nutr 71:142–151.

Nguyen U, Yuqing Z, Yanyan Z, Jingbo N, Bin Z, Felson D. 2011. Increasing prevalence of knee pain and symptomatic knee osteoarthritis: survey and cohort data. Ann Intern Med 55:725–732.

Nicklisch N, Maixner F, Ganslmeier R, Friederich S, Dresely V, Meller H, Zink A, Alt KW. 2012. Rib lesions in skeletons from early Neolithic sites in central Germany: on the trail of tuberculosis at the onset of agriculture. Am J Phys Anthropol 149:391–404.

Nilsson O, Baron J. 2004. Fundamental limits on longitudinal bone growth: growth plate senescence and epiphyseal fusion. Trends Endocrinol Metab 15:370–374.

Niu J, Zhang Y, Torner J, Nevitt M, Lewis CE, Aliabadi P, Sack B, Clancy M, Sharma L, Felson D. 2009. Is obesity a risk factor for progressive radiographic knee osteoarthritis? Arthritis Rheum 61:329–335.

Normando D, Faber J, Guerreiro JF, Quintão CCA. 2011. Dental occlusion in a split Amazon indigenous population: genetics prevails over environment. PLoS One 6:e28387.

Novak M, Šlaus M. 2011. Vertebral pathologies in two early modern period (16th–19th century) populations from Croatia. Am J Phys Anthropol 145:270–281.

Nowak O, Piontek J. 2002. The frequency of appearance of transverse (Harris) lines in the tibia in relationship to age at death. Ann Hum Biol 29:314–325.

Nunn JH. 1996. Prevalence of dental erosion and the implications for oral health. Eur J Oral Sci 104:156–161.

Nyazee HA, Finney KM, Sarikonda M, Towler DA, Johnson JE, Babcock HM. 2012. Diabetic foot osteomyelitis: bone markers and treatment outcomes. Diabetes Res Clin Pract 97:411–417.

Oakley Jr GP. 2009. The scientific basis for eliminating folic acid–preventable spina bifida: a modern miracle from epidemiology. Ann Epidemiol 19:226–230.

O'Brien E, Jorde LB, Rönnlöf B, Fellman JO, Eriksson AW. 1988. Founder effect and genetic disease in Sottunga, Finland. Am J Phys Anthropol 77:335–346.

Ochoa-Acuña H, Carbajo C. 2009. Risk of limb birth defects and mother's home proximity to cornfields. Sci Total Environ 407:4447–4451.

O'Connor S, Ali E, Al-Sabah S, Anwar D, Bergström E, Brown KA, Buckberry J, Buckley S, Collins M, Denton J, et al. 2011. Exceptional preservation of a prehistoric human brain from Heslington, Yorkshire, UK. J Archaeol Sci 38:1641–1654.

Oddy WH, De Klerk NH, Miller M, Payne J, Bower C. 2009. Association of maternal pre-pregnancy weight with birth defects: evidence from a case-control study in Western Australia. Aust NZ J Obstet Gynaecol 49:11–15.

Ojima M, Hanioka T, Tanaka K, Aoyama H. 2007. Cigarette smoking and tooth loss experience among young adults: a national record linkage study. BMC Public Health 7:313.

Olds TS, Harten NR. 2001. One hundred years of growth: the evolution of height, mass, and body composition in Australian children, 1899–1999. Hum Biol 73:727–738.

O'Leary CM, Bower C, Knuiman M, Stanley FJ. 2007. Changing risks of stillbirth and neonatal mortality associated with maternal age in Western Australia 1984–2003. Paediatr Perinat Epidemiol 21:541–549.

Olivieri F, Semproli S, Pettener D, Toselli S. 2008. Growth and malnutrition of rural Zimbabwean children (6–17 years of age). Am J Phys Anthropol 136:214–222.

O'Neill BJ, Molloy AP, McCarthy T. 2013. Osteomyelitis of the tibia following anterior cruciate ligament reconstruction. Int J Surg Case Rep 4:143–145.

Ong G, Yeo JF, Bhole S. 1996. A survey of reasons for extraction of permanent teeth in Singapore. Community Dent Oral Epidemiol 24:124–127.

Ong KK, Ahmed ML, Emmett PM, Preece MA, Dunger DB. 2000. Association between postnatal catch-up growth and obesity in childhood: prospective cohort study. BMJ 320:967–971.

Ortner DJ. 1968. Description and classification of degenerative bone changes in the distal surfaces of the humerus. Am J Phys Anthropol 28:139–156.

Östberg AL, Nyholm M, Gullberg B, Råstam L, Lindblad U. 2009. Tooth loss and obesity in a defined Swedish population. Scand J Public Health 37:427–433.

Oxenham MF, Matsumura H. 2008. Oral and physiological paleohealth in cold adapted peoples: Northeast Asia, Hokkaido. Am J Phys Anthropol 135:64–74.

Oxenham MF, Thuy NK, Cuong NL. 2005. Skeletal evidence for the emergence of infectious disease in Bronze and Iron Age northern Vietnam. Am J Phys Anthropol 126:359–376.

Oxenham MF, Tilley L, Matsumura H, Nguyen LC, Nguyen KT, Nguyen KD, Domett K, Huffer D. 2009. Paralysis and severe disability requiring intensive care in Neolithic Asia. Anthropol Sci 117:107–112.

Paans N, van den Akker-Scheek I, Dilling RG, Bos M, van der Meer K, Bulstra SK, Stevens M. 2013. Effect of exercise and weight loss in people who have hip osteoarthritis and are overweight or obese: a prospective cohort study. Phys Ther 93:137–146.

Pace E. 1997 Oct 30. T. Dale Stewart dies at 96; anthropologist at Smithsonian. New York Times.

Page CJ, Hinman RS, Bennell KL. 2011. Physiotherapy management of knee osteoarthritis. Int J Rheum Dis 14:145–151.

Page WF, Hoaglund FT, Steinbach LS, Heath AC. 2003. Primary osteoarthritis of the hip in monozygotic and dizygotic male twins. Twin Res 6:147–151.

Palmert MR, Mansfield MJ, Crowley WF, Crigler JF, Crawford JD, Boepple PA. 1999. Is obesity an outcome of gonadotropin-releasing hormone agonist administration? Analysis of growth and body composition in 110 patients with central precocious puberty. J Clin Endocrinol Metab 84:4480–4488.

Palubeckaitė-Miliauskienė Z, Jankauskas R. 2007. Dental status of two military samples: soldiers of Napoleon's great army and German soldiers in World War I. Papers on Anthropology 16:222–236.

Pany D, Teschler-Nicola M. 2007. Klippel-Feil syndrome in an early Hungarian period juvenile skeleton from Austria. Int J Osteoarchaeol 17:403–415.

Papageorgopoulou C, Suter SK, Rühli FJ, Siegmund F. 2011. Harris lines revisited: prevalence, comorbidities, and possible etiologies. Am J Hum Biol 23:381–391.

Parfitt AM. 2002. Misconceptions (1): epiphyseal fusion causes cessation of growth. Bone 30:337–339.

Parker SE, Werler MM, Shaw GM, Anderka M, Yazdy MM. 2012. Dietary glycemic index and the risk of birth defects. Am J Epidemiol 176:1110–1120.

Parker-Pope T. 2000 Mar 10. A common side effect, dry mouth, can cause serious tooth decay. Wall Street Journal. B1.

Pearson KL. 1997. Nutrition and the early-medieval diet. Speculum 72:1–32.

Peckmann T. 2003. Possible relationship between porotic hyperostosis and smallpox infections in nineteenth-century populations in the northern frontier, South Africa. World Archaeol 35:289–305.

Peltonen M, Lindroos A, Torgerson J. 2003. Musculoskeletal pain in the obese: a comparison with a general population and long-term changes after conventional and surgical obesity treatment. Pain 104:549–557.

Peng B, Chen J, Kuang Z, Li D, Pang X, Zhang X. 2009. Diagnosis and surgical treatment of back pain originating from endplate. Eur Spine J 18:1035–1040.

Peng B, Wu W, Hou S, Shang W, Wang X, Yang Y. 2003. The pathogenesis of Schmorl's nodes. J Bone Joint Surg Br 85:879–882.

Peres MA, Tsakos G, Barbato PR, Silva DA, Peres KG. 2012. Tooth loss is associated with increased blood pressure in adults—a multidisciplinary population-based study. J Clin Periodontol 39:824–833.

Perry MA. 2005. Redefining childhood through bioarchaeology: toward an archaeological and biological understanding of children in antiquity. Archeological Papers of the American Anthropological Association 15:89–111.

Perzigian AJ. 1973. Osteoporotic bone loss in two prehistoric Indian populations. Am J Phys Anthropol 39:87–95.

Petersen PE, Ogawa H. 2012. The global burden of periodontal disease: towards integration with chronic disease prevention and control. Periodontology 2000 60:15–39.

Pētersone-Gordina E, Gerhards G. 2011. Dental disease in a 17th–18th century German community in Jelgava, Latvia. Papers on Anthropology 20:327–350.

Pförtner J, Hövel M. 2003. Bony ghosts—residual effects of severe growth arrest. N Engl J Med 349:18.

Phillips SM, Sivilich M. 2006. Cleft palate: a case study of disability and survival in prehistoric North America. Int J Osteoarchaeol 16:528–535.

Pinhasi R, Shaw P, White B, Ogden AR. 2006. Morbidity, rickets and long-bone growth in post-medieval Britain—a cross-population analysis. Ann Hum Biol 33:372–389.

Pitre MC, Lovell NC. 2010. A sacral anomaly from the Quaker cemetery, Kingston-upon-Thames, England. Int J Osteoarchaeol 20:351–357.

Plomp KA, Roberts CA, Viðarsdóttir US. 2012. Vertebral morphology influences the development of Schmorl's nodes in the lower thoracic vertebrae. Am J Phys Anthropol 149:572–582.

Pocock NA, Eisman JA, Hopper JL, Yeates MG, Sambrook PN, Eberl S. 1987. Genetic determinants of bone mass in adults. A twin study. J Clin Invest 80:706–710.

Pollack P. 2008. Impact of childhood obesity on bones. AAOS Now [Internet]. [accessed November 6, 2013]; 2(9).

Pollock NK, Bernard PJ, Gutin B, Davis CL, Zhu H, Dong Y. 2011. Adolescent obesity, bone mass, and cardiometabolic risk factors. J Pediatr 158:727–734.

Polzer I, Schwahn C, Völzke H, Mundt T, Biffar R. 2012. The association of tooth loss with all-cause and circulatory mortality. Is there a benefit of replaced teeth? A systematic review and meta-analysis. Clin Oral Investig 16:333–351.

Poulsen LW, Qvesel D, Brixen K, Vesterby A, Boldsen JL. 2001. Low bone mineral density in the femoral neck of medieval women: a result of multiparity? Bone 28:454–458.

Prasad KC, Sreedharan S, Chakravarthy Y, Prasad SC. 2007. Tuberculosis in the head and neck: experience in India. J Laryngol Otol 121:979–985.

Pratt CA, Hekmat M, Barnard JD, Zaki GA. 1998. Indications for third molar surgery. J Roy Coll Surg Edinb 43:105–108.

Prentice A, Schoenmakers I, Laskey MA, de Bono S, Ginty F, Goldberg GR. 2006. Nutrition and bone growth and development. Proc Nutr Soc 65:348–360.

Pretty GL, Henneberg M, Lambert KM, Prokopec M. 1998. Trends in stature in the South Australian Aboriginal Murraylands. Am J Phys Anthropol 106:505–514.

Promislow JH, Goodman-Gruen D, Slymen DJ, Barrett-Connor E. 2002. Protein consumption and bone mineral density in the elderly the Rancho Bernardo study. Am J Epidemiol 155:636–644.

Psoter WJ, Reid BC, Katz RV. 2005. Malnutrition and dental caries: a review of the literature. Caries Res 39:441–447.

Punzi L, Oliviero F, Plebani M. 2005. New biochemical insights into the pathogenesis of osteoarthritis and the role of laboratory investigations in clinical assessment. Crit Rev Clin Lab Sci 42:279–309.

Quesnele J, Dufton J, Stern P. 2012. Spinal infection: a case report. J Can Chiropr Assoc 56:209–215.

Raff J, Cook DC, Kaestle F. 2006. Tuberculosis in the New World: a study of ribs from the Schild Mississippian population, west-central Illinois. Mem Inst Oswaldo Cruz 101:25–27.

Ragle RL, Sawitzke AD. 2012. Nutraceuticals in the management of osteoarthritis. Drugs Aging 29:717–731.

Ramegowda S, Ramachandra NB. 2006. Parental consanguinity increases congenital heart diseases in south India. Ann Hum Biol 33:519–528.

Rankin J, Tennant PWG, Stothard KJ, Bythell M, Summerbell CD, Bell R. 2010. Maternal body mass index and congenital anomaly risk: a cohort study. Int J Obes 34:1371–1380.

Rankin-Hill LM, Blakey ML. 1994. W. Montague Cobb (1904–1990): physical anthropologist, anatomist, and activist. Am Anthropol 96:74–96.

Rapuri PB, Gallagher JC, Kinyamu HK, Ryschon KL. 2001. Caffeine intake increases the rate of bone loss in elderly women and interacts with vitamin D receptor genotypes. Am J Clin Nutr 74:694–700.

Rasmussen SA, Chu SY, Kim SY, Schmid CH, Lau J. 2008. Maternal obesity and risk of neural tube defects: a meta-analysis. Am J Obstet Gynecol 198:611–619.

Rauch F, Travers R, Munns C, Glorieux FH. 2004. Sclerotic metaphyseal lines in a child treated with pamidronate: histomorphometric analysis. J Bone Miner Res 19:1191–1193.

Ravn P, Cizza G, Bjarnason NH, Thompson D, Daley M, Wasnich RD, McClung M, Hosking D, Yates AJ, Christiansen C. 1999. Low body mass index is an important risk factor for low bone mass and increased bone loss in early postmenopausal women. J Bone Miner Res 14:1622–1627.

Reece EA. 2008. Obesity, diabetes, and links to congenital defects: a review of the evidence and recommendations for intervention. J Matern Fetal Neonatal Med 21:173–180.

Reefhuis J, Honein MA, Schieve LA, Correa A, Hobbs CA, Rasmussen SA. 2009. Assisted reproductive technology and major structural birth defects in the United States. Hum Reprod 24:360–366.

Reginster JY, Neuprez A, Lecart MP, Sarlet N, Bruyere O. 2012. Role of glucosamine in the treatment for osteoarthritis. Rheumatol Int 32:2959–2967.

Reijman M, Hazes JMW, Pols HAP, Koes BW, Bierma-Zeinstra SMA. 2005. Acetabular dysplasia predicts incident osteoarthritis of the hip. Arthritis Rheum 52:787–793.

Reinhard KJ, Tieszen L, Sandness KL, Beiningen LM, Miller E, Ghazi A, Miewalk CD, Barnum SV. 1994. Trade, contact, and female health in northeast Nebraska. In: Larsen CS, Milner GJ, editors. In the wake of contact: Biological responses to conquest. New York: Wiley-Liss. p. 63–74.

Reitman CA, Gertzbein SD, Francis Jr WR. 2002. Lumbar isthmic defects in teenagers resulting from stress fractures. Spine J 2:303–306.

Rice JE, Skull SA, Pearce C, Mulholland N, Davie G, Carapetis JR. 2003. Screening for intestinal parasites in recently arrived children from East Africa. J Paediatr Child Health 39:456–459.

Richards GD, Antón SC. 1991. Craniofacial configuration and postcranial development of a hydrocephalic child (ca. 2500 BC–500 AD): with a review of cases and comment on diagnostic criteria. Am J Phys Anthropol 85:185–200.

Rick DJ, Rees CA, Osborn KA, Crookston BT, Leaver K, Merrill SB, Velásquez C, Ricks JH. 2012. Peru's national folic acid fortification program and its effect on neural tube defects in Lima. Rev Panam Salud Publica 32:391–398.

Rick TC, Erlandson JM, Vellanoweth RL, Braje TJ. 2005. From Pleistocene mariners to complex hunter-gatherers: The archaeology of the California Channel Islands. J World Prehist 19:169–228.

Ridgewell M. 2003. Managing back pain. Update 67:263–269.

Rinaldi A. 2012. Yaws eradication: facing old problems, raising new hopes. PLoS Negl Trop Dis 6:e1837.

Risser WL. 1991. Weight-training injuries in children and adolescents. Am Fam Physician 44:2104–2108.

Rivera JA, Hotz C, González-Cossío T, Neufeld L, García-Guerra A. 2003. The effect of micronutrient deficiencies on child growth: a review of results from community-based supplementation trials. J Nutr 133:4010S–4020S.

Rizzoli R, Bonjour JP. 2004. Dietary protein and bone health. J Bone Miner Res 19:527–531.

Roberts CA. 2000. Infectious disease in biocultural perspective: past, present and future work in Britain. In: Cox M, Mays S, editors. Human osteology in archaeology and forensic science. London: Greenwich Medical Media. p. 145–162.

Roberts CA, Knüsel CJ, Race L. 2004. A foot deformity from a Romano-British cemetery at Gloucester, England, and the current evidence for talipes in palaeopathology. Int J Osteoarchaeol 14:389–403.

Robins AH. 2009. The evolution of light skin color: role of vitamin D disputed. Am J Phys Anthropol 139:447–450.

Rod-Fleury T, Dunkel N, Assal M, Rohner P, Tahintzi P, Bernard L, Hoffmeyer P, Lew D, Uçkay I. 2011. Duration of post-surgical antibiotic therapy for adult chronic osteomyelitis: a single-centre experience. Int Orthop 35:1725–1731.

Rogers J, Shepstone L, Dieppe P. 1997. Bone formers: osteophytes and enthesophyte formation are positively associated. Ann Rheum Dis 56:85–90.

Rogers J, Shepstone L, Dieppe P. 2004. Is OA a systemic disorder of bone? Arthritis Rheum 50:452–457.

Rogers J, Waldron T, Watt I. 1991. Erosive osteoarthritis in a medieval skeleton. Int J Osteoarchaeol 1:151–153.

Rojas-Sepúlveda C, Ardagna Y, Dutour O. 2008. Paleoepidemiology of vertebral degenerative disease in a pre-Columbian Muisca series from Colombia. Am J Phys Anthropol 135:416–430.

Rosen CJ, Bouxsein ML. 2006. Mechanisms of disease: is osteoporosis the obesity of bone? Nat Clin Pract Rheumatol 2:35–43.

Rosin AJ. 1970. Clinical features and detection of osteomalacia in the elderly. Postgrad Med J 46:131–136.

Rossignol M. 2004. Osteoarthritis and occupation in the Quebec national health and social survey. Occup Environ Med 61:729–735.

Rothschild BM. 1997. Porosity: a curiosity without diagnostic significance. Am J Phys Anthropol 104:529–533.

Rothschild BM. 2012. Extirpation of the mythology that porotic hyperostosis is caused by iron deficiency secondary to dietary shift to maize. Adv Anthropol 2:157–160.

Rothschild BM, Calderon FL, Coppa A, Rothschild C. 2000. First European exposure to syphilis: the Dominican Republic at the time of Columbian contact. Clin Infect Dis 31:936–941.

Rothschild BM, Woods RJ. 1991a. Spondyloarthropathy: erosive arthritis in representative defleshed bones. Am J Phys Anthropol 85:125–134.

Rothschild BM, Woods RJ. 1991b. Arthritis in an early 20th century geriatric population. Age 14:17–19.

Rothschild BM, Woods RJ, Ortel W. 1990. Rheumatoid arthritis "in the buff": erosive arthritis in defleshed bones. Am J Phys Anthropol 82:441–449.

Rubin DM, Christian CW, Bilaniuk LT, Zazyczny KA, Durbin DR. 2003. Occult head injury in high-risk abused children. Pediatrics 111:1382–1386.

Ruff C, Holt B, Trinkaus E. 2006. Who's afraid of the big bad Wolff? "Wolff is law" and bone functional adaptation. Am J Phys Anthropol 129:484–498.

Rühli FJ, Henneberg M. 2002. Are hyperostosis frontalis interna and leptin linked? A hypothetical approach about hormonal influence on human microevolution. Med Hypotheses 58:378–381.

Ruiz-Cotorro A, Balius-Matas R, Estruch-Massana A, Vilará AJ. 2006. Spondylolysis in young tennis players. Br J Sports Med 40:441–446.

Rumball JS, Lebrun CM, Di Ciacca SR, Orlando K. 2005. Rowing injuries. Sports Med 35:537–555.

Russell SL, Gordon S, Lukacs JR, Kaste LM. 2013. Sex/gender differences in tooth loss and edentulism: historical perspectives, biological factors, and sociologic reasons. Dent Clin North Am 57:317–337.

Sadeghpour A, Nakamoto T. 2011. WIPO Patent Application PCT/US2011/024734.

Sairyo K, Sakai T, Yasui N, Kiapour A, Biyani A, Ebraheim N, Goel VK. 2009. Newly occurred L4 spondylolysis in the lumbar spine with pre-existence L5 spondylolysis among sports players: case reports and biomechanical analysis. Arch Orthop Trauma Surg 129:1433–1439.

Sajko S, Stuber K, Wessely M. 2011. Growth restart/recovery lines involving the vertebral body: a rare, incidental finding and diagnostic challenge in two patients. J Can Chiropr Assoc 55:313–317.

Sambrook PN, MacGregor AJ, Spector TD. 1999. Genetic influences on cervical and lumbar disc degeneration: a magnetic resonance imaging study in twins. Arthritis Rheum 42:366–372.

Sanders M, Grundmann O. 2011. The use of glucosamine, devil's claw (*Harpagophytum procumbens*), and acupuncture as complementary and alternative treatments for osteoarthritis. Altern Med Rev 16:228–238.

Sandhu J, Ben-Shlomo Y, Cole TJ, Holly J, Smith GD. 2006. The impact of childhood body mass index on timing of puberty, adult stature and obesity: a follow-up study based on adolescent anthropometry recorded at Christ's Hospital (1936–1964). Int J Obes 30:14–22.

Sandmark H, Jogstedt C, Vingård E. 2000. Primary osteoarthrosis of the knee in men and women as a result of lifelong physical load from work. Scand J Work Environ Health 26:20–25.

Sankar WN, Horn BD, Wells L, Dormans JP. 2011. Slipped capital femoral epiphysis. In: Kliegman RM, Behrman RE, Jenson HB, Stanton BF, editors. Nelson textbook of pediatrics. 19th ed. Philadelphia (PA): Saunders Elsevier. p. 6363–6364.

Santos AL, Roberts CA. 2001. A picture of tuberculosis in young Portuguese people in the early 20th century: a multidisciplinary study of the skeletal and historical evidence. Am J Phys Anthropol 115:38–49.

Santos AL, Suby JA. 2012. Skeletal and surgical evidence for acute osteomyelitis in non-adult individuals. Int J Osteoarchaeol. doi:10.1002/oa.2276.

Sarov B, Bentov Y, Kordysh E, Karakis I, Bolotin A, Hershkovitz R, Belmaker I. 2008. Perinatal mortality and residential proximity to an industrial park. Arch Environ Occup Health 63:17–25.

Sartoris DJ, Resnick D, Tyson R, Haghighi P. 1985. Age-related alterations in the vertebral spinous processes and intervening soft tissues: radiologic-pathologic correlation. AJR Am J Roentgenol 145:1025–1030.

Sasa T, Yoshizumi Y, Imada K, Aoki M, Terai T, Koizumi T, Goel VK, Faizan A, Biyani A, Sakai T, Sairyo K. 2009. Cervical spondylolysis in a judo player: a case report and biomechanical analysis. Arch Orthop Trauma Surg 129:559–567.

Schaefer K, Lauc T, Mitteroecker P, Gunz P, Bookstein FL. 2006. Dental arch asymmetry in an isolated Adriatic community. Am J Phys Anthropol 129:132–142.

Schaible UE, Kaufmann SHE. 2007. Malnutrition and infection: complex mechanisms and global impacts. PLoS Med 4:e115.

Schermer SJ. 2013 Jan 15. E-mail message to author.

Schmitt H, Brocai DR, Lukoschek M. 2004. High prevalence of hip arthrosis in former elite javelin throwers and high jumpers: 41 athletes examined more than 10 years after retirement from competitive sports. Acta Orthop Scand 75:34–39.

Schmitt JV, Dechandt IT, Dopke G, Ribas ML, Cerci FB, Viesi JMZ, Marchioro HZ, Zunino MMB, Miot HA. 2010. Armadillo meat intake was not associated with leprosy in a case control study, Curitiba (Brazil). Mem Inst Oswaldo Cruz 105:857–862.

Scholes D, Hubbard RA, Ichikawa LE, LaCroix AZ, Spangler L, Beasley JM, Reed S, Ott SM. 2011. Oral contraceptive use and bone density change in adolescent and young adult women: a prospective study of age, hormone dose, and discontinuation. J Clin Endocrinol Metab 96:E1380–E1387.

Schreider E. 1967. Body-height and inbreeding in France. Am J Phys Anthropol 26:1–3.

Schultz JJ. 1998. A comparison of osteoarthritis in the appendicular joints of individuals from the Hamann-Todd and Terry collections [MA thesis]. University of Indianapolis.

Schwager T. 2010 July/August. Adolescent growth plate injuries: how to protect, diagnose, and treat young athletes. American Fitness. 24–27.

Seck BC, Jackson RT. 2010. Multiple contributors to iron deficiency and anemia in Senegal. Int J Food Sci Nutr 61:204–216.

Seeman E. 2002. Pathogenesis of bone fragility in women and men. Lancet 359:1841–1850.

Seki S, Yikawaguchi Y, Chiba K, Mikami Y, Kizawa H, Oya T, Mio F, Mori M, Miyamoto Y, Masuda I, et al. 2005. A functional SNP in CILP, encoding cartilage intermediate layer protein is associated with susceptibility to lumbar disc disease. Nat Genet 37:607–612.

Serrano M, Han M, Brinez P, Linask KK. 2010. Fetal alcohol syndrome: cardiac birth defects in mice and prevention with folate. Am J Obstet Gynecol 203:75.e7–75.e15.

Shaffer JR, Wang X, DeSensi RS, Wendell S, Weyant RJ, Cuenco KT, Crout R, McNeil DW, Marazita ML. 2012. Genetic susceptibility to dental caries on pit and fissure and smooth surfaces. Caries Res 46:38–46.

Shalitin S, Phillip M. 2003. Role of obesity and leptin in the pubertal process and pubertal growth—a review. Int J Obes 27:869–874.

Shawky RM, Sadik DI. 2011. Congenital malformations prevalent among Egyptian children and associated risk factors. Egyptian Journal of Medical Human Genetics 12:69–78.

Shepard GJ, Banks AJ, Ryan WG. 2003. Ex-professional association footballers have an increased prevalence of osteoarthritis of the hip compared with age matched controls despite not having sustained notable hip injuries. Br J Sports Med 37:80–81.

Sherman JD. 1996. Chlorpyrifos (Dursban)-associated birth defects: report of four cases. Arch Environ Health 51:5–8.

Shin DH, Oh CS, Kim YS, Hwang YI. 2012. Ancient-to-modern secular changes in Korean stature. Am J Phys Anthropol 147:433–442.

Shrier I. 2001. Spondylolysis incidence in various sports. Physical Sports Medicine 29:5.

Shuler KA. 2011. Life and death on a Barbadian sugar plantation: historic and bioarchaeological views of infection and mortality at Newton Plantation. Int J Osteoarchaeol 21:66–81.

Sigurdsson G, Halldorsson BV, Styrkarsdottir U, Kristjansson K, Stefansson KS. 2008. Impact of genetics on low bone mass in adults. J Bone Miner Res 23:1584–1590.

Silventoinen K, Kaprio J, Yokoyama Y. 2011. Genetics of pre-pubertal growth: a longitudinal study of Japanese twins. Ann Hum Biol 38:608–614.

Skinner M, McLaren M, Carlson RL. 1988. Therapeutic cauterization of periodontal abscesses in a prehistoric Northwest coast woman. Med Anthropol Q 2:278–285.

Šlaus M. 2000. Biocultural analysis of sex differences in mortality profiles and stress levels in the late medieval population from Nova Rača, Croatia. Am J Phys Anthropol 111:193–209.

Šlaus M. 2008. Osteological and dental markers of health in the transition from the late antique to the early medieval period in Croatia. Am J Phys Anthropol 136:455–469.

Slipman CW, Shin CH, Patel RK, Isaac Z, Huston CW, Lipetz JS, Lenrow DA, Braverman DL, Vresilovic EJ. 2002. Etiologies of failed back surgery syndrome. Pain Med 3:200–214.

Slon V, Nagar Y, Kuperman T, Hershkovitz I. 2011. A case of dwarfism from the Byzantine city Rehovot-in-the-Negev, Israel. Int J Osteoarchaeol 23:573–589.

Smith MO. 2006. Treponemal disease in the middle Archaic to early Woodland periods of the western Tennessee River Valley. Am J Phys Anthropol 131:205–217.

Smith MO, Dorsz JR, Betsinger TK. 2013. Diffuse idiopathic skeletal hyperostosis (DISH) in pre-Columbian North America: evidence from the eastern Tennessee River Valley. Int J Paleopath 3:11–18.

Smithsonian National Museum of Natural History [SNMNH], Physical Anthropology Collections. http://anthropology.si.edu/cm/phys_intro.htm.

Solomons NW. 2003. Environmental contamination and chronic inflammation influence human growth potential. J Nutr 133:1237–1237.

Sonne-Holm S, Jacobsen S, Rovsing HC, Monrad H, Gebuhr P. 2007. Lumbar spondylolysis: a lifelong dynamic condition? A cross sectional survey of 4.151 adults. Eur Spine J 16:821–828.

Sowers M, Zobel D, Hawthorne VM, Carmen W, Weissfeld L. 1991. Progression of OA of the hand and metacarpal bone loss. A twenty-year follow up of incident cases. Arthritis Rheum 34:36–42.

Specker B, Binkley T. 2005. High parity is associated with increased bone size and strength. Osteoporosis Int 16:1969–1974.

Spector TD, Cicuttini F, Baker J, Loughlin J, Hart D. 1996. Genetic influences on osteoarthritis in women: a twin study. BMJ 312:940–944.

Spector TD, MacGregor AJ. 2004. Risk factors for osteoarthritis: genetics. Osteoarthritis Cartilage 12:S39–S44.

Spellberg B, Bartlett JG, Gilbert DN. 2013. The future of antibiotics and resistance. N Engl J Med 368:299–302.

Spellberg B, Lipsky BA. 2012. Systemic antibiotic therapy for chronic osteomyelitis in adults. Clin Infect Dis 54:393–407.

Steele JG, Treasure E, Pitts NB, Morris J, Bradnock G. 2000. Adult dental health survey: total tooth loss in the United Kingdom in 1998 and implications for the future. Br Dent J 189:598–603.

Steffian AF, Simon JJ. 1994. Metabolic stress among prehistoric foragers of the central Alaskan Gulf. Arctic Anthropol 31:78–94.

Steyn M. 2011. Case report: forensic anthropological assessment in a suspected case of child abuse from South Africa. Forensic Sci Int 208:e6–e9.

Stinchfield K. 2008. Little athletes, big injuries. Time 171:51.

Stini WA. 1990. "Osteoporosis": etiologies, prevention, and treatment. Yrbk Phys Anthropol 33:151–194.

Stoll C, Alembik Y, Dott B, Roth MP. 1992. An epidemiologic study of environmental and genetic factors in congenital hydrocephalus. Eur J Epidemiol 8:797–803.

Streeten EA, Ryan KA, McBride DJ, Pollin TI, Shuldiner AR, Mitchell BD. 2005. The relationship between parity and bone mineral density in women characterized by a homogeneous lifestyle and high parity. J Clin Endocrinol Metab 90:4536–4541.

Strömqvist B. 2002. Evidence-based lumbar spine surgery: the role of national registration. Acta Orthop 73:34–39.

Styne DM. 2004. Puberty, obesity and ethnicity. Trends Endocrinol Metab 15: 472–478.

Sulosky Weaver CL, Wilson RJA. 2012. Probable atretic cephalocele in an adult female from Punta Secca (Sicily, Italy). Int J Osteoarchaeol. doi:10.1002/oa.2291.

Sun L, Zhang L, Wang K, Wang W, Tian M. 2012. Fungal osteomyelitis after arthroscopic anterior cruciate ligament reconstruction: a case report with review of the literature. The Knee 19:728–731.

Sunder M. 2008. Shrinking due to corpulence? BMI in childhood predicts subsequent linear growth among US children and youth, 1963–1970. Ann Hum Biol 35:432–438.

Sutton RA, Mumm S, Coburn SP, Ericson KL, Whyte MP. 2012. "Atypical femoral fractures" during bisphosphonate exposure in adult hypophosphatasia. J Bone Miner Res 27:987–994.

Suzuki T, Fujita H, Choi JG. 2008. Brief communication: new evidence of tuberculosis from prehistoric Korea—population movement and early evidence of tuberculosis in Far East Asia. Am J Phys Anthropol 136:357–360.

Suzuki T, Inoue T. 2007. Earliest evidence of spinal tuberculosis from the Aneolithic Yayoi period in Japan. Int J Osteoarchaeol 17:392–402.

Swanston T, Carter Y, Hopkins C, Walker EG, Cooper DML. 2011. Developmental fusion of the malleus and incus in a late 19th-century case of aural atresia. Int J Osteoarchaeol 23:612–617.

Sylvester AD, Christensen AM, Kramer PA. 2006. Factors influencing osteological changes in the hands and fingers of rock climbers. J Anat 209:597–609.

Szabó N, Gergev G, Valek A, Eller J, Kaizer L, Sztriha L. 2013. Birth prevalence of neural tube defects: a population-based study in south-eastern Hungary. Childs Nerv Syst 29:621–627.

Taes Y, Lapauw B, Vanbillemont G, Bogaert V, De Bacquer D, Goemaere S, Zmierczak H, Kaufman J-M. 2010. Early smoking is associated with peak bone mass and prevalent fractures in young, healthy men. J Bone Miner Res 25:379–387.

Tan KA, Walker M, Morris K, Greig I, Mason JI, Sharpe RM. 2006. Infant feeding with soy formula milk: effects on puberty progression, reproductive function and testicular cell numbers in marmoset monkeys in adulthood. Hum Reprod 21:896–904.

Tayles N, Buckley HR. 2004. Leprosy and tuberculosis in Iron Age Southeast Asia? Am J Phys Anthropol 125:239–256.

Teebi AS, El-Shanti HI. 2006. Consanguinity: implications for practice, research, and policy. Lancet 367:970–971.

Temple DH. 2008. What can variation in stature reveal about environmental differences between prehistoric Jomon foragers? Understanding the impact of systemic stress on developmental stability. Am J Hum Biol 20:431–439.

Temple DH, Larsen CS. 2007. Dental caries prevalence as evidence for agriculture and subsistence variation during the Yayoi period in prehistoric Japan: biocultural interpretations of an economy in transition. Am J Phys Anthropol 134:501–512.

Ten Kate LP, Rath M, Felbor U, Frints SM. 2013. Birth defects after incestuous mating: calculating the probability of causality and reflecting on the desirability of genetic testing. Eur J Med Genet 56:243–244.

Teotia SPS, Teotia M. 2008. Nutritional bone disease in Indian population. Indian J Med Res 127:219–228.

Texas State University Department of Anthropology, Forensic Anthropology Center [FACTS]. http://www.txstate.edu/anthropology/facts.

Thomas T, Burguera B. 2002. Is leptin the link between fat and bone mass? J Bone Miner Res 17:1563–1569.

Thompson AR. 2012. Differential diagnosis of limb length discrepancy in a 19th century burial from southwest Mississippi. Int J Osteoarchaeol. doi:10.1002/oa.2238.

Thomson WM. 2005. Issues in the epidemiological investigation of dry mouth. Gerodontology 22:65–76.

Tickle M, Moulding G, Milsom K, Blinkhorn A. 2000. Dental public health: dental caries, contact with dental services and deprivation in young children: their relationship at a small area level. Br Dent J 189:376–379.

Trembly DL. 1995. On the antiquity of leprosy in western Micronesia. Int J Osteoarchaeol 5:377–384.

Trinkaus E, Zimmerman MR. 1982. Trauma among Shanidar Neanderthals. Am J Phys Anthropol 57:61–76.

Trotter M, Gleser GC. 1951. Trends in stature of American whites and Negroes born between 1840 and 1924. Am J Phys Anthropol 9:427–440.

Truswell AS. 1985. ABC of nutrition. Other nutritional deficiencies in affluent communities. BMJ 291:1333–1337.

Turner RT. 2000. Skeletal response to alcohol. Alcohol Clin Exp Res 24:1693–1701.

Ubelaker DH, Pap I. 2009. Skeletal evidence for morbidity and mortality in Copper Age samples from northeastern Hungary. Int J Osteoarchaeol 19:23–35.

Umeta M, West CE, Verhoef H, Haidar J, Hautvast JG. 2003. Factors associated with stunting in infants aged 5–11 months in the Dodota-Sire District, rural Ethiopia. J Nutr 133:1064–1069.

University of New Mexico [UNM] Laboratory of Human Osteology, Maxwell Museum of Anthropology Documented Skeletal Collection. http://www.unm.edu/~osteolab/coll_doc.html.

University of Tennessee, Knoxville [UTK], Forensic Anthropology Center, WM Bass Donated Skeletal Collection. http://web.utk.edu/~fac/collection.html.

Unnanuntana A, Toogood P, Hart D, Cooperman D, Grant RE. 2010. Evaluation of proximal femoral geometry using digital photographs. J Orthop Res 28:1399–1404.

Upex BR, Knüsel CJ. 2009. Avulsion fractures of the transverse processes of the first thoracic vertebra: an archaeological case study from Raunds. Int J Osteoarchaeol 19:116–122.

US Census Bureau. http://www.census.gov.

Üstündağ H. 2009. Schmorl's nodes in a post-medieval skeletal sample from Klostermarienberg, Austria. Int J Osteoarchaeol 19:695–710.

Valdes AM, Hart DJ, Jones KA, Surdulescu G, Swarbrick P, Doyle DV, Schafer AJ, Spector TD. 2004. Association study of candidate genes for the prevalence and progression of knee osteoarthritis. Arthritis Rheum 50:2497–2507.

van der Heijden C, Claerbout M, Peers K, Lauweryns P. 2007. Traumatic lumbar pedicle fracture associated with pre-existing bilateral spondylolysis and anterolisthesis in a professional soccer player. J Back Musculoskelet Rehabil 20:131–134.

van der Merwe AE, Maat GJR, Steyn M. 2010. Ossified haematomas and infectious bone changes on the anterior tibia: histomorphological features as an aid for accurate diagnosis. Int J Osteoarchaeol 20:227–239.

van der Pal-de Bruin KM, van der Heijden PG, Buitendijk SE, den Ouden AL. 2003. Periconceptional folic acid use and the prevalence of neural tube defects in the Netherlands. Eur J Obstet Gynecol Reprod Biol 108:33–39.

van der Westhuizen J, Mennen U. 2010. A working classification for the management of scapho-trapezium-trapezoidosteo-arthritis. Hand Surg 15:203–210.

van Doren JM, Kleinmeier D, Hammack TS, Westerman A. 2013. Prevalence, serotype diversity, and antimicrobial resistance of Salmonella in imported shipments of spice offered for entry to the United States, FY2007–FY2009. Food Microbiol 34:239–251.

van Saase JL, van Romunde LK, Cats A, Vandenbroucke JP, Valkenburg HA. 1989. Epidemiology of osteoarthritis: Zoetermeer survey. Comparison of radiological osteoarthritis in a Dutch population with that in 10 other populations. Ann Rheum Dis 48:271–280.

Vandana KL, Reddy S. 2007. Assessment of periodontal status in dental fluorosis subjects using community periodontal index of treatment needs. Indian J Dent Res 18:67–71.

Veselka B, Hoogland MLP, Waters-Rist AL. 2012. Rural rickets: Beemster, a rural farming community in post-medieval Netherlands [MSc thesis]. University of Leiden.

Vialle R, Ilharreborde B, Dauzac C, Lenoir T, Rillardon L, Guigui P. 2007. Is there a sagittal imbalance of the spine in isthmic spondylolisthesis? A correlation study. Eur Spine J 16:1641–1649.

Victora CG, Barros FC, Horta BL, Martorell R. 2001. Short-term benefits of catch-up growth for small-for-gestational-age infants. Int J Epidemiol 30:1325–1330.

Vinceti M, Rovesti S, Bergomi M, Calzolari E, Candela S, Campagna A, Milan M, Vivoli G. 2001. Risk of birth defects in a population exposed to environmental lead pollution. Sci Total Environ 278:23–30.

Vodanović M, Peroš K, Zukanović A, Knežević M, Novak M, Šlaus M, Brkić H. 2012. Periodontal diseases at the transition from the late antique to the early mediaeval period in Croatia. Arch Oral Biol 57:1362–1376.

von Hunnius TE, Yang D, Eng B, Waye JS, Saunders SR. 2007. Digging deeper into the limits of ancient DNA research on syphilis. J Archaeol Sci 34:2091–2100.

Wakai K, Naito M, Naito T, Kojima M, Nakagaki H, Umemura O, Yokota M, Hanada N, Kawamura T. 2010. Tooth loss and intakes of nutrients and foods: a nationwide survey of Japanese dentists. Community Dent Oral Epidemiol 38:43–49.

Waldron T. 1997. Osteoarthritis of the hip in past populations. Int J Osteoarchaeol 7:186–189.

Waldvogel FA. 2004. Infectious diseases in the 21st century: old challenges and new opportunities. Int J Infect Dis 8:5–12.

Walker D. 2009. The treatment of leprosy in 19th century London: a case study from St. Marylebone cemetery. Int J Osteoarchaeol 19:364–374.

Walker PL 1983. Is the battered child syndrome a modern phenomenon. In Proceedings of the Tenth European Meeting of the Paleopathology Association.

Walker PL, Bathurst RR, Richman R, Gjerdrum T, Andrushko VA. 2009. The causes of porotic hyperostosis and cribra orbitalia: a reappraisal of the iron-deficiency-anemia hypothesis. Am J Phys Anthropol 139:109–125.

Walker PL, Cook DC, Lambert PM. 1997. Skeletal evidence for child abuse: a physical anthropological perspective. J Forensic Sci 42:196–207.

Walker R, Hill K, Burger O, Hurtado AM. 2006. Life in the slow lane revisited: ontogenetic separation between chimpanzees and humans. Am J Phys Anthropol 129:577–583.

Wandera M, Engebretsen IM, Okullo I, Tumwine JK, Åstrøm AN. 2009. Socio-demographic factors related to periodontal status and tooth loss of pregnant women in Mbale district, Uganda. BMC Oral Health 9:18.

Waner S, Durrheim DN, Leggat PA, Ross MH. 2001. Preventing infectious diseases in long-term travelers to rural Africa. J Travel Med 8:304–308.

Wang Q, Alén M, Nicholson P, Lyytikäinen A, Suuriniemi M, Helkala E, Suominen H, Cheng S. 2005. Growth patterns at distal radius and tibial shaft in pubertal girls: a 2-year longitudinal study. J Bone Miner Res 20:954–961.

Wang SC, Wang L, Lee MC. 2012a. Adolescent mothers and older mothers: who is at higher risk for adverse birth outcomes? Public Health 126:1038–1043.

Wang TK, Wong CF, Au WK, Cheng VC, Wong SS. 2007. *Mycobacterium tuberculosis* sternal wound infection after open heart surgery: a case report and review of the literature. Diagn Microbiol Infect Dis 58:245–249.

Wang X, Willing MC, Marazita ML, Wendell S, Warren JJ, Broffitt B, Smith B, Busch T, Lidral AC, Levy SM. 2012b. Genetic and environmental factors associated with dental caries in children: the Iowa fluoride study. Caries Res 46:177–184.

Ward CV, Mays SA, Child S, Latimer B. 2010. Lumbar vertebral morphology and isthmic spondylolysis in a British medieval population. Am J Phys Anthropol 141:273–280.

Watkins IV RG, Watkins III RG. 2010. Lumbar spondylolysis and spondylolisthesis in athletes. Semin Spine Surg 22:210–217.

Watson JT. 2008a. Changes in food processing and occlusal dental wear during the early agricultural period in northwest Mexico. Am J Phys Anthropol 135:92–99.

Watson JT. 2008b. Prehistoric dental disease and the dietary shift from cactus to cultigens in northwest Mexico. Int J Osteoarchaeol 18:202–212.

Watson JT, Fields M, Martin DL. 2010. Introduction of agriculture and its effects on women's oral health. Am J Human Biol 22:92–102.

Watts R. 2011. Non-specific indicators of stress and their association with age at death in medieval York: using stature and vertebral neural canal size to examine the effects of stress occurring during different periods of development. Int J Osteoarchaeol 21:568–576.

Wedman J, van Weissenbruch R. 2005. Chronic recurrent multifocal osteomyelitis. Ann Otol Rhinol Laryngol 114:65–68.

Weedon MN, Frayling TM. 2008. Reaching new heights: insights into the genetics of human stature. Trends Genet 24:595–603.

Wegehaupt FJ, Günthart N, Sener B, Attin T. 2011. Prevention of erosive/abrasive enamel wear due to orange juice modified with dietary supplements. Oral Dis 17:508–514.

Wei S, Jones G, Thomson R, Dwyer T, Venn A. 2011. Oral contraceptive use and bone mass in women aged 26–36 years. Osteoporosis Int 22:351–355.

Weisberg SP, McCann D, Desai M, Rosenbaum M, Leibel RL, Ferrante Jr AW. 2003. Obesity is associated with macrophage accumulation in adipose tissue. J Clinic Invest 112:1796–1808.

Weise M, De-Levi S, Barnes KM, Gafni RI, Abad V, Baron J. 2001. Effects of estrogen on growth plate senescence and epiphyseal fusion. Proc Natl Acad Sci USA 98:6871–6876.

Weiss E. 2005. Schmorl's nodes: A preliminary investigation. Paleopathology Newsletter 132:6–10.

Weiss E. 2006. Osteoarthritis and body mass. J Archaeol Sci 33:690–695.

Weiss E. 2008. When it rains it pours: multiple congenital pathologies in single individuals. Am J Phy Anthropol 46S:220.

Weiss E. 2009. Spondylolysis in a pre-contact San Francisco Bay population: behavioral and anatomical sex differences. Int J Osteoarchaeol 19:375–385.

Weiss E. 2013. Hand OA and bone loss: is there an inverse relationship? HOMO 64:357–365.

Weiss E, Jurmain RD. 2007. Osteoarthritis revisited: a contemporary review of aetiology. Int J Osteoarchaeol 17:437–450.

Wekselman K, Spiering K, Hetteberg C, Kenner C, Flandermeyer A. 1995. Fetal alcohol syndrome from infancy through childhood: a review of the literature. J Pediatr Nurs 10:296–303.

Wells C. 1962. Joint pathology in ancient Anglo-Saxons. J Bone Joint Surg 44(B):948.

Welten DC, Kemper HC, Bertheke Post G, Staveren WAV. 1995. A meta-analysis of the effect of calcium intake on bone mass in young and middle aged females and males. J Nutr 125:2802–2813.

Wendell S, Wang X, Brown M, Cooper ME, DeSensi RS, Weyant RJ, Crout R, McNeil DW, Marazita ML. 2010. Taste genes associated with dental caries. J Dent Res 89:1198–1202.

Wentz RK, De Grummond NT. 2009. Life on horseback: palaeopathology of two Scythian skeletons from Alexandropol, Ukraine. Int J Osteoarchaeol 19:107–115.

Wentzel P. 2009. Can we prevent diabetic birth defects with micronutrients? Diabetes Obes Metab 11:770–778.

Weston DA. 2008. Investigating the specificity of periosteal reactions in pathology museum specimens. Am J Phys Anthropol 137:48–59.

Wheeler SM, Williams L, Beauchesne P, Dupras TL. 2013. Shattered lives and broken childhoods: evidence of physical child abuse in ancient Egypt. Int J Paleopath. doi:10.1016/j.ijpp.2013.03.009.

White CD, Armelagos GJ. 1997. Osteopenia and stable isotope ratios in bone collagen of Nubian female mummies. Am J Phys Anthropol 103:185–199.

White T, Folkens P. 1991. Human osteology. San Diego: Academic Press.

Whitehead RG. 1977. Protein and energy requirements of young children living in the developing countries to allow for catch-up growth after infections. Am J Clinic Nutr 30:1545–1547.

Whittaker DK, Molleson T, Nuttall T. 1998. Calculus deposits and bone loss on the teeth of Romano-British and eighteenth-century Londoners. Arch Oral Biol 43:941–948.

Wicker A. 2008. Spondylolysis and spondylolisthesis in sports—FIMS position statement. Int SportMed J 9:74–78.

Widdowson EM, Oftedal OT, Whiten A, Ulijaszek SJ. 1991. Contemporary human diets and their relation to health and growth: overview and conclusions [and discussion]. Philos Trans R Soc Lond B Biol Sci 334:289–295.

Wiegand A, Burkhard JPM, Eggmann F, Attin T. 2012. Brushing force of manual and sonic toothbrushes affects dental hard tissue abrasion. Clin Oral Investig 17:815–822.

Wiklund PK, Xu L, Wang Q, Mikkola T, Lyytikäinen A, Völgyi E, Munukka E, Cheng SM, Alen M, Keinänen-Kiukaanniemi S, Cheng S. 2012. Lactation is associated with greater maternal bone size and bone strength later in life. Osteoporosis Int 23:1939–1945.

Wilczak C, Mulhern D. 2012. Co-occurrence of DISH and HFI in the Terry Collection. Int J Osteoarchaeol 22:452–459.

Williams C, Sutcliffe A, Sebire NJ. 2010. Congenital malformations after assisted reproduction: risks and implications for prenatal diagnosis and fetal medicine. Ultrasound Obstet Gynecol 35:255–259.

Williams FMK, Manek NJ, Sambrook PN, Spector TD, Macgregor AJ. 2007. Schmorl's nodes: common, highly heritable, and related to lumbar disc disease. Arth Care Res 57:855–860.

Williams JPG. 1981. Catch-up growth. J Embryol Exp Morphol Supp 65:89–101.

Williams NR. 2003. Joint replacement. Update 67:215–220.

Willis TA. 1923. The lumbo-sacral vertebral column in man, its stability of form and function. Am J Anat 32:95–123.

Wilson DJ. 2012. Insights from genomics into bacterial pathogen populations. PLoS Pathog 8:e1002874.

Wilson MG, Michet CJ, Illstrup DM, Melton LJ. 1990. Idiopathic symptomatic osteoarthritis of the hip and knee: a population-based incidence study. Mayo Clin Proc 65:1214–1221.

Wolff J. 1986. The law of bone remodelling. Berlin: Springer-Verlag.

Wolff MS, Teitelbaum SL, Pinney SM, Windham G, Liao L, Biro F, Kushi LH, Erdmann C, Hiatt RA, Rybak ME, Calafat AM. 2010. Investigation of relationships between urinary biomarkers of phytoestrogens, phthalates, and phenols and pubertal stages in girls. Environ Health Perspect 118:1039–1046.

Woo SL, Kuei SC, Amiel D, Gomez MA, Hayes WC, White FC, Akeson WH. 1981. The effect of prolonged physical training on the properties of long bone: a study of Wolff's law. J Bone Joint Surg Am 63A:780–786.

Wood JW, Milner GR, Harpending HC, Weiss KM. 1992. The osteological paradox: problems of inferring prehistoric health from skeletal samples. Curr Anthropol 33:343–370.

Woods RJ. 1995. Biomechanics of osteoarthritis of the knee [PhD dissertation]. Ohio State University.

Wördemann M, Polman K, Menocal Heredia LT, Junco Diaz R, Collado Madurga A-M, Núñez Fernández FA, Cordovi Prado RA, Ruiz Espinosa A, Pelayo Duran L, Bonet Gorbea M, Rojas Rivero L, Gryseels B. 2006. Prevalence and risk factors of intestinal parasites in Cuban children. Trop Med Int Health 11:1813–1820.

World Health Organization [WHO]. http://www.who.int/en.

Wright LE, Chew F. 1998. Porotic hyperostosis and paleoepidemiology: a forensic perspective on anemia among the ancient Maya. Am Anthropol 100:924–939.

Wu DY, Brat G, Milla G, Kim J. 2007. Knowledge and use of folic acid for prevention of birth defects amongst Honduran women. Reprod Toxicol 23:600–606.

Wu J, Chen G, Liao Y, Song X, Pei L, Wang J, Zheng X. 2011. Arsenic levels in the soil and risk of birth defects: a population-based case-control study using GIS technology. J Environ Health 74:20–25.

Wu SH, Ho CT, Nah SL, Chau CF. 2014. Global hunger: a challenge to agricultural, food, and nutritional sciences. Crit Rev Food Sci Nutr 54:151–162.

Yahata Y, Aoyagi K, Yoshida S, Ross PD, Yoshimi I, Moji K, Takemoto T. 2002. Appendicular bone mass and knee and hand OA in Japanese women, a cross-sectional study. BMC Musculoskelet Disord 3:24–34.

Yamazaki T, Yamori M, Yamamoto K, Saito K, Asai K, Sumi E, Goto K, Takahashi K, Nakayama T, Bessho K. 2012. Risk of osteomyelitis of the jaw induced by oral bisphosphonates in patients taking medications for osteoporosis: a hospital-based cohort study in Japan. Bone 51:882–887.

Yan JH, Gu WJ, Sun J, Zhang WX, Li BW, Pan L. 2013. Efficacy of tai chi on pain, stiffness and function in patients with osteoarthritis: a meta-analysis. PLoS One 8:e61672.

Yang S, Jawahar R, McAlindon TE, Eaton CB, Lapane KL. 2012a. Racial differences in symptom management approaches among persons with radiographic knee osteoarthritis. BMC Complement Altern Med 12:86.

Yang W, Zeng L, Cheng Y, Chen Z, Wang X, Li X, Yan H. 2012b. The effects of periconceptional risk factor exposure and micronutrient supplementation on birth defects in Shaanxi Province in western China. PLoS One 7:e53429.

Yüksel Ş, Kutlubay A, Karaoğlu L, Yoloğlu S. 2009. The prevalence of consanguineous marriages in the city of Malatya, Turkey. Turkish Journal of Medical Sciences 39:133–137.

Zaki ME, Hussien FH, El Banna RAES. 2009. Osteoporosis among ancient Egyptians. Int J Osteoarchaeol 19:78–89.

Zhai G, Hart DJ, Kato BS, MacGregor A, Spector TD. 2007. Genetic influence on the progression of radiographic knee osteoarthritis: a longitudinal twin study. Osteoarthritis Cartilage 15:222–225.

Zhang Y, Hunter DJ, Nevitt MC, Xu L, Niu J, Lui LY, Wei Y, Aliabadi P, Felson DT. 2004. Association of squatting with increased prevalence of radiographic tibiofemoral knee osteoarthritis: the Beijing Osteoarthritis Study. Arthritis Rheum 50:1187–1192.

Zias J, Mitchell P. 1996. Psoriatic arthritis in a fifth-century Judean desert monastery. Am J Phys Anthropol 101:491–502.

Zink AR, Molnar E, Motamedi N, Palfy G, Marcsik A, Nerlich AG. 2007. Molecular history of tuberculosis from ancient mummies and skeletons. Int J Osteoarchaeol 17:380–391.

Zinoviev AV. 2010. Review of human osseal remains from XVI–XVIII centuries cemetery of Zatveretsky Posad (Tver, Russia). Bull Int Assoc Paleodont 4:11–21.

Zollikofer CP, Ponce de León MS. 2010. The evolution of hominin ontogenies. Semin Cell Dev Biol 21:441–452.

Zumwalt AC. 2006. The effect of endurance exercise on the morphology of muscle attachment sites. J Exp Biol 209:444–454.

Index

abrasion, 107, 109–12
abscess, 117–18, 117*f*
acetabular dysplasia. *See* dysplasia
activity. *See* exercise
acupuncture, 14, 103
adipocytes, 30; and menstruation, 53; and osteoporosis, 5
age: and archeological record, 9; and back pain, 73; and child abuse, 61–62; and menarche, 30–32, 53, 102; and menopause, 43, 45, 47; and osteoarthritis, 96–97; and osteoporosis, 42–43, 45–46; parental, 151, 164–67, 178–79; and periodontal disease, 118; and tooth loss, 122; and tooth wear, 110. *See also* life expectancy
agriculture: and infections, 130; and osteomyelitis, 136; and periodontal disease, 118; and stature, 28; and tooth decay, 113–14
alcohol: and birth defects, 151, 167–68; and bone health, 50–51; and Harris lines, 24
allele(s), 44; and bone mineral density, 44; and osteoarthritis, 96; and vitamin D, 52
alternative medicine: ethnicity and, 14; for osteoarthritis, 103. *See also* acupuncture; chiropractic care

alveolar bone, 107; bone loss in, 116; leprosy and, 148
amelia, 158–59
amenorrhea, 46; in female athletes, 53
ancient DNA (aDNA), 130, 142, 146
anemia, 19, 21, 25, 88, 131, 176, 178; food fads and, 180; parasite induced, 137–40. *See also* iron deficiency
anencephaly, 153, 169–71
ankylosing spondylitis, 87–88
anterolisthesis, 82
antibiotic resistance, 126, 132, 136, 147, 150
apes, 17–18, 71–72, 105
arsenic, 162
artificial reproductive technologies (ART), 165–66, 178–79; multiple births, effects of, 166–67, 179. *See also* pregnancy
arthritis, 8; erosive, 87–88; psoriatic, 88; rheumatoid, 88; treatment of, 103–4
athletes. *See* sports
attrition, 107, 109–12, 109*f*, 127; and agriculture, 113–14; and calculus, 118
autism, 151, 179
autoimmune diseases, 88, 98
autopsy collections, 10–13; and arthritis, 93, 95–97; and bone mass, 46; and congenital defects, 153; and fractures, 39; issues with, 13–14, 27; and

hyperglycemia, and congenital defects, 169
hyperostosis frontalis interna, 101, 101*f*, 102–3

inbreeding: and birth defects, 163–64, 176; and malocclusion, 126; and spina bifida, 153; and supernumerary teeth, 161
Industrial Age, 6–7; malocclusion, 125; stature, 29; urbanization, 176
infectious diseases, 6, 129–50; eradication of, 142–43, 148–50; return of, 149, 179–80
in-toeing, 69
in vitro fertilization (IVF), 166
iron deficiency, 21–22, 137–40, 177. *See also* anemia

joint pain. *See* arthritis

Kennewick Man, 9
kidney disease, and osteomalacia, 38
Klippel-Feil syndrome, 152, 155
kwashiorkor, 23
kyphosis, 23, 74, 83, 144–45, 151–152

lactation, 43; and osteoporosis, 46–49
lateral epicondylar fracture, 69–70, 69*f*
lead, 115, 123, 162, 180
leprosy, 130, 143, 147–49, 152, 159
leptin(s), 5; hyperostosis frontalis interna and, 102–103; inflammation and, 100; puberty and, 30, 53
life expectancy, 13, 46, 174
lithium, and osteoporosis, 42
Little League elbow, 64
Looser's zone, 37
lordosis, 74, 80, 83
lumbarization, 80, 81*f*, 84

magnesium, 20; deficiency of, and teeth, 108; vegetarians and, 50
malnutrition, 6, 8, 20–23, 33, 117, 146, 151, 177–78; and Harris lines, 24–25;

return of, 179–80; and stature, 26–29, 173. *See also* nutrition
malocclusion, 107, 124–26, 125*f*, 168; congenital diseases and, 168; genetics of, 125; in the Hamann-Todd Collection, 125; and molar impaction, 126; temporal changes in, 124–25; in the Terry Collection, 125
Marfan syndrome, 151
measles, 149, 179
medieval period: and attrition, 110; and bone loss, 46–47, 51, 180; and Harris lines, 24–25; and hydrocephaly, 156; and infections, 132, 138; and malocclusion, 124; and osteomalacia, 38; and rickets, 22; and stature, 28; and trauma, 60, 77; and vertebral fractures, 53
menarche, 18; age of, trends in, 30–32, 102; and bone mineral density, 43, 45; female athletes and, 53
menopause, 47, 51; and hyperostosis frontalis interna, 102; and osteoporosis, 42–43, 45
mercury, 180
meromelia, 158
methicillin-resistant *Staphylococcus aureus* (MRSA), 14, 136, 150
Middle Ages: menarche in the, 30; and mycobacterial diseases, 143–48
military records, and stature, 29
molars, 18, 107–8, 110; abscess of, 117*f*; extraction of, 126; necrosis, 126
muscle, pulled, 56, 66–67
mycobacterial diseases, 143–150

Neanderthals, 6, 120, 124
Neolithic, 6; caries rates in the, 113; menarche in the, 30; tooth pain relief in the, 111
neural tube defects, 151–53, 162, 166–67; folic acid fortification and, 169–71, 177–78; obesity and, 168–69; skull anomalies, 156–58
neuropathy, 136

About the Author

Elizabeth Weiss is professor of anthropology at San José State University in California, where, since 2004, she has been teaching physical anthropology courses including Introduction to Human Evolution, Human Origins, Primates, Mummies, Human Osteology, Human Variation, Modernity and Disease, and Bioarchaeology. She received her PhD from the University of Arkansas.

Professor Weiss's research expertise is in skeletal analyses of osteoarthritis, muscle markers, and bone cross-sections to reconstruct lifestyles and better understand bone biology. She uses data on skeletal remains and big clinical data sets to place health and disease in perspective.

Weiss has presented research at annual meetings of the American Association of Physical Anthropologists, the Canadian Association of Physical Anthropology, the Paleopathology Association, the American Anthropological Association, the American Association for the Advancement of Science, and the Center for Healthy Aging in Multicultural Populations. Some of her presentation titles have included "Is Seventy the New Forty? Paleopathology and Aging," "Calcaneal Enthesophytes: Etiology beyond Activity," and "A Pain in the Neck: Caudal Shifting Occipitalization in a California Amerind Adult."

Along with three books and one anthology, Weiss has published over three dozen articles, chapters, opinion pieces, and book reviews in top anthropology journals (such as the *American Journal of Physical Anthropology* and the *International Journal of Osteoarchaeology*) and medical journals (such as *Rheumatology*), with topics ranging from analysis of 2.5-million-year-old *Homo habilis* foot fossils to the relationship among knee osteoarthritis, body weight, and pain. Her research has been featured in *Science*, Reuters, and the History News Network, and elsewhere in the media.